WHICH COMES FIRST,
CARDIO
OR
WEIGHTS?

Alex Hutchinson, Ph.D.

WHICH COMES FIRST,

CARDIO

OR

WEIGHTS?

Fitness Myths,
Training Truths, and Other
Surprising Discoveries from
the Science of Exercise

HARPER

NEW YORK • LONDON • TORONTO • SYDNEY

HARPER

HarperCollins books may be purchased for educational, business, or sales promotional use. For information, please write: Special Markets Department, HarperCollins Publishers, 10 East 53rd Street, New York, NY 10022.

FIRST EDITION

Designed by Leah Springate

Library of Congress Cataloging-in-Publication Data

Hutchinson, Alex.
Which comes first, cardio or weights? / Alex Hutchinson.—1st ed.
    p. cm.

Includes bibliographical references.

ISBN 978-0-06-200753-7

1. Physical fitness. 2. Exercise. 3. Health. I. Title.

RA781.H798 2011
613.7—dc22

                                                    2010053598

This book is for informational purposes only. Consult your health care provider before beginning any exercise plan. The author and the publisher expressly disclaim responsibility for any adverse effects arising from the use or application of the information contained herein.

Illustrations by Trish McAlaster

11 12 13 14 15   10 9 8 7 6 5 4 3 2 1

For Lauren

CONTENTS

Introduction: What This Book Is (and Isn't)

WHEN A RESEARCH INSTITUTE in La Jolla, California, issued a press release with the provocative title "Exercise in a Pill" in the summer of 2008, it wasn't hard to predict the headlines that would be splashed across the next day's newspapers. "Trying to reap the health benefits of exercise? Forget treadmills and spin classes," the release proclaimed. "Researchers at the Salk Institute for Biological Studies may have found a way around the sweat and pain." Feeding a drug called AICAR to a group of mice for four weeks allowed them to run 44 percent longer than drug-free mice, the new study found—without a single step of training.

Stories like this are hard to resist. After all, who wouldn't love a pain-free, effortless shortcut to fitness? That's why we're bombarded on a daily basis by similar promises in magazines, on television, over the Internet, and in our spam folders. There's just one problem with these get-fit-quick schemes: they don't work.

Exercise, as medical physiologists Frank Booth and Matthew Laye of the University of Missouri pointed out in a critique of the Salk Institute study, affects almost every organ system in your body—circulatory, neural, endocrine, gastrointestinal, immune, kidney, skeletal muscle, bone, ligament, and so on. In fact, the only systems that don't respond to exercise appear to be the senses and (surprisingly) the lungs. Unless a pill can

1

somehow produce changes in all these systems simultaneously, it will never replace exercise. (Moreover, you don't get healthy by having the *capacity* to run 44 percent farther—you can only burn calories and get the resulting benefits by actually *doing* it.)

So let me begin with a full disclosure: this book does not contain any secret workouts or magic pills that will produce instant fitness.

Instead, what follows is an up-to-date guide to what scientists know about exercise, health, and performance—and, just as importantly, what they don't know. It's an "evidence-based" guide: the answers it offers to common fitness questions aren't based on conventional wisdom or gut feelings. Instead, they're drawn from the more than 400 peer-reviewed journal articles listed in the reference section at the back, along with over 100 interviews with researchers around the world.

That means that if scientists don't yet have a definitive answer to a question, I haven't just made one up. For example, the last few years have seen a surge of interest in running barefoot or in "minimalist" shoes, a practice that some people believe reduces the risk of injury compared with conventional running shoes. A few studies examining how running barefoot affects your joints have now been published, but the results so far have been ambiguous—so, on page 38, you can read about some of these studies and make up your own mind about whether the potential benefits outweigh the risks (and the funny looks). For someone who has been repeatedly injured trying to run in conventional shoes, the answer might be yes; for someone trying running for the first time, probably not.

This is an important point: there's no single "best" exercise program or technique that applies to everyone. You'll have to take into account your background, current level of fitness, and goals in designing an appropriate workout regimen—not to mention more subtle considerations like the types of activity you enjoy. After all, the most effective program is the one you can stick with!

No matter what type of exercise you choose, there are some basic concepts that govern the training process, in particular, the "Specific Adaptation to Imposed Demands" (SAID) principle:

When the body is subjected to stresses and overloads of varying intensities, it will gradually adapt over time to overcome whatever demands are placed on it.

That's what every form of exercise boils down to. The stress could be lifting weights or pedaling a bicycle, and the adaptation is bigger muscle fibers, a stronger heart, and hundreds of other microscopic changes. The key is balancing the size of the stress: too small (lifting a half-pound weight, say), and your body won't see any need to adapt; too large, and it won't have a chance to adapt due to injury or exhaustion. Much of the research described in this book aims to help you find this delicate balance.

Your body's overall goal in making these adaptations is, simply put, to more efficiently convert the energy from your food into physical action. Every move you make is powered by the contraction of muscle fibers, and these contractions are fueled by a molecule called ATP (adenosine triphosphate) extracted from the carbohydrate, fat, and protein you've eaten.

Your body has several different ways of converting nutrients to ATP. One supply is stored in a form called phosphocreatine, which can provide energy for short, intense bursts lasting up to about 10 seconds. A process called glycolysis offers another source of ATP that lasts about 45 seconds. Neither of these processes requires oxygen to create ATP, so they're known as "anaerobic" (literally "without oxygen") energy sources. For longer spurts of exercise, the tiny "power plants" in your cells called mitochondria can convert carbohydrate or fat to ATP, but the process requires oxygen—which is why sustained exercise is referred to as "aerobic."

As your body adapts to exercise, you'll become more efficient at every step of this process. Your cells generate more mitochondria, enabling you to produce more ATP; you start burning a higher proportion of fat (a virtually unlimited form of stored energy) instead of carbohydrate; and the muscle fibers themselves produce more force with each contraction.

In addition to these functional adaptations, your body will remodel its structure. Less fat and bigger muscles are the two most obvious changes, but other adjustments will be taking place under the surface. The constant tug of stronger muscles on your skeleton results in stronger bones; your heart literally grows bigger in order to pump more oxygen-rich blood to hard-working muscles; you develop a more extensive network of capillaries to distribute that blood; the nerves that carry commands from your brain to your muscles learn to do so more quickly and efficiently; and on and on.

In some ways, though, the most important adaptations— and perhaps the most unexpected—are those that occur in the brain. This is a field of research that remains in its infancy, though it has been gathering momentum for the past few years. But what we do know for sure is that the increased flow of blood and growth factors to the brain during exercise has dramatic effects, boosting memory and learning, enhancing cognition, and warding off the effects of aging. And exercise also stimulates the release of powerful mood-altering chemicals like endorphins, so much so that some researchers argue that it can become a mild addiction.

If that's the case, it's an outcome that I devoutly wish for everyone who reads this book. When, as a teenager, I started trying to run on a daily basis, it was a tremendous chore. It stayed that way for several years—until, during a layoff due to injury, I suddenly realized how much I missed it. These days, my workout is the highlight of my day, a chance to get outside, unwind, perhaps spend time with my wife or friends, or simply be alone with my thoughts. I don't believe that researchers will ever succeed in producing "exercise in a pill"—but if they did, I wouldn't be interested in taking it. I hope that, after reading this book and putting some of its ideas into practice, you'll end up feeling the same way.

1: Getting Started

In 1960, President-Elect John F. Kennedy wrote an article for *Sports Illustrated* called "The Soft American," lamenting the declining role of physical activity in everyday life. "Today human activity, the labor of the human body, is rapidly being engineered out of working life," he wrote. "By the 1970s, according to many economists, the man who works with his hands will be almost extinct." That prediction didn't quite pan out—but it's certainly true that, for most of us, physical activity is now a choice rather than the necessity it was for our ancestors.

Kennedy was concerned that Americans wouldn't have the "vigor and determination" necessary to match the Soviet Union; today, we've realized that physical fitness is essential to the health of our bodies and minds. But the basic challenge remains the same: if you've just begun working out recently, or you're about to head to the gym for the first time, you need to know what to expect. How hard to push, how long it will take to see results, how to make the most of your workout time, how to minimize the risks associated with starting an exercise program—these are the issues you should consider before getting started.

How long does it take to get in shape?

First, the good news. Your body actually starts getting stronger and healthier just hours after you start working out. But if you're wondering how long it will take to rock a six-pack—well, you'll have to be a bit more patient. A few years ago, exercise scientist Megan Anderson and her colleagues at the University of Wisconsin-La Crosse put 25 sedentary volunteers through an intense six-week exercise program modeled on the bold claims made by companies such as Bowflex and Body-for-Life. Despite sticking to the program religiously, zero percent of the subjects developed instant washboard abs. In fact, a panel of six judges could detect no differences whatsoever in their physical attractiveness before and after the program.

That doesn't mean nothing was happening. After just a few strength training sessions, your brain learns to recruit more muscle fibers and make them contract all at once to produce a greater force. This "neural activation" kicks in after only a few workouts, allowing you to get stronger almost immediately, well before your muscles get noticeably bigger. Further strength gains come as the individual muscle fibers within your muscles get bigger, which starts in as little as two weeks if you're training intensely. But it takes longer for these changes to be noticeable: even with sophisticated lab equipment, researchers can't usually detect changes in fat and muscle composition until after about nine weeks of training. Similarly, a University of Tokyo study published in 2010 saw the biggest increases in strength after two months and the biggest boost in muscle size after three months. To achieve these rapid gains, the subjects were doing four very hard workouts a week. For the average person at the gym, it will take six months or more to see significant sculpting of the body—even though strength has been increasing from day one.

Weight loss is more difficult to predict, because it depends on your starting point, your health history, your genetics, and your diet as well as your workout routine. But like strength training, aerobic exercise produces major health and performance benefits long before you see them in the mirror. Aerobic exercise increases the number of mitochondria, which are essentially the "cellular power plants" in your muscles that use oxygen to produce energy: the more mitochondria you have, the farther and faster you can run, and the more fat your muscles will burn. Studies have found that about six weeks of training will boost mitochondria levels by 50 to 100 percent.

Health benefits, on the other hand, kick in after a single bout of aerobic exercise. For about 48 hours after a workout, your muscles will be consuming more glucose than usual, helping to bring down blood-sugar levels. After a few workouts, your insulin sensitivity will begin to improve, offering further control of blood sugar.

The bottom line: "Getting in shape" is a journey that extends over months and even years, but the process—and the benefits—start as soon as you begin exercising. So if you're having trouble staying motivated without immediate physical changes, track your strength and endurance gains along the way.

Am I exercising enough?

This is a hugely controversial question, but not in the way you might expect. Almost all experts agree about what the science says—but they're bitterly divided about what message they should convey to the public. Decades of research have made two things crystal clear:

1. Every bit of exercise helps, even in scraps as short as 10 minutes.
2. More is almost always better.

The challenge is conveying the second message without discouraging the people who are still struggling with the first. And there are a lot of people struggling: in 2008, the Centers for Disease Control and Prevention found that a quarter of Americans hadn't done any physical activity at all in the previous month. And less than half met the modest government-mandated Healthy People 2010 goal of 30 minutes of moderate exercise five times a week. In Canada, only a third of people are meeting a similar goal of 30 minutes at a moderate effort, four times a week.

From a public health perspective, the top priority is getting those inactive people to start moving, even a little bit. Going from zero to slightly active offers the biggest possible health boost, according to a recent National Institutes of Health study that followed 250,000 men and women between the ages of 50 and 71. Those who were just slightly active but didn't manage to meet the exercise guidelines were 30 percent less likely to die than those who were totally inactive. Stepping it up to moderate exercise reduced risk by only eight more percent, and adding in some vigorous exercise subtracted an additional 12 percent. When you combine that data, you find that getting half an hour of moderate to vigorous exercise five times a week cuts your risk of dying from all causes in half. So far, so good. The controversy starts when you ask what happens if you do more exercise than the government guidelines recommend. According to researchers like Paul Williams of the Lawrence Berkeley National Laboratory, you keep piling up more and more benefits. Williams has been following a cohort of 120,000 runners since 1991, tracking how much they run and what happens to their health. With this enormous sample size, his National Runners' Health Study has been able to uncover a pronounced "dose–response" relationship between aerobic exercise and health: the more you do, and the harder you do it, the more benefits you get.

What kind of benefits, you ask? Well, in a series of studies stretching back over a decade, Williams has found that the risk of everything from big killers like diabetes, stroke, and heart attack to less common conditions like glaucoma, cataracts, and macular degeneration can be reduced by as much as 70 percent by going beyond the standard exercise guidelines. In every case, the benefits relate to both how much running the subjects do and how fast they do it. For example, every additional mile in your average daily run lowers your glaucoma risk by 8 percent. And speeding up your 10K time by one meter per second (the difference between, say, finishing times of 53 minutes and 40 minutes) lowers heart attack risk by about 50 percent.

It is true that overdoing your workout regimen can weaken your immune system (see p. 23), especially if you're not eating and sleeping well. And elite athletes training for hours a day at extremely high intensities sometime suffer from "overtraining" syndrome. But the message of Williams's research is that the risks of doing too much—and these results apply to all forms of aerobic exercise, not just running—pale in comparison to the benefits of doing a bit more.

Which should I do first: cardio or weights?

Let's start with one incontrovertible fact: you can't fulfill your ultimate potential as both a weightlifter and a marathoner at the same time. Too many hours sweating on the elliptical will hinder your ability to put on muscle, and pumping too much iron will slow your endurance gains. But most of us don't want

Olympic medals in both events. We just want some combination of reasonable cardiovascular fitness and non-vanishing muscles—a desire shared by many elite athletes. Top basketball players, for instance, need strength and explosiveness but also have to last for a full 40 to 60 minutes on the court.

The solution, according to Derek Hansen, the head coach for strength and conditioning at Simon Fraser University in Vancouver, British Columbia, and a speed consultant to numerous Olympic athletes, is to mix it up. For basketball players, he says, "we typically have our athletes lift weights, jump, and sprint one day, then do their aerobic work the next day." When Hansen's court-sport athletes are combining weight training with cardio in a single session, the weights come first, since building power is their first priority.

This approach—starting with whichever activity is most important to you—is widely used by elite athletes. Until recently, scientists thought it was simply a matter of logistics: if you're tired from the treadmill, you can't lift as much weight, so over time you put on less muscle. But new techniques now allow researchers to directly measure which specific proteins are produced in muscles after different types of exercises. It turns out that the sequence of cellular events that leads to bigger muscles is determined in part by the same "master switch"—an enzyme called AMP kinase—that controls adaptations for better endurance. But you can't have it both ways: the switch is set either to "bigger muscles" or to "better endurance," and the body can't instantly change from one setting to the other. How you start your workout determines which way the switch will be set for the session.

So if your goal is beach muscles, your weights routine should come first. If you're preparing for an upcoming 5K race, do your full cardio workout before tacking on weights at the end. And if you're looking for the best of both worlds, Hansen suggests mixing it up, both within a single session and from day to day: "The variability will be good, as it challenges your body and metabolism."

Can I get fit in seven minutes a week?

Breathless claims about exercise regimens that produce near-instant results with minimal effort are generally the domain of late-night infomercials. So it might seem surprising that one of the hot topics at the American College of Sports Medicine's annual meeting over the last few years has been research into "high-intensity interval training" (HIT), whose proponents suggest that many of the benefits of traditional endurance training can be achieved with a few short bouts of intense exercise totaling as little as seven minutes a week.

Exercise physiologist Martin Gibala and his colleagues at McMaster University in Hamilton, Ontario, have performed a remarkable series of studies in which their subjects cycle as hard as they can for 30 seconds, then rest for four minutes, and repeat four to six times. They do this short workout three times a week. "The gains are quite substantial," Gibala says. Compared to control subjects who cycle continuously for up to an hour a day, five times a week, the HIT subjects show similar gains in exercise capacity, muscle metabolism, and cardiovascular fitness. In fact, the group's latest study shows that HIT improves the structure and function of key arteries that deliver blood to the muscles and heart—just like typical cardio training. Similar studies by University of Guelph researcher Jason Talanian have found that high-intensity interval training also increases the body's ability to burn fat, an effect that persists even during lower-intensity activities following the interval training.

The results are no surprise to elite cyclists, runners, and swimmers, who have relied on interval training for decades to achieve peak performance. To break the four-minute mile in 1954, Roger Bannister famously relied on interval sessions of ten 60-second sprints separated by two minutes of rest, because his duties as a medical student on clinical rotation limited his training time to half an hour a day at lunch. Such time constraints are the main reason Gibala advocates HIT, since studies consistently find that lack of time is the top reason that people don't manage to get the 30 minutes of daily exercise recommended by public health guidelines. "We're not saying that it's a panacea that has all the benefits of endurance training," he says. "But it's a way that people can get away with less."

More recent studies have started to piece together exactly how HIT sessions work. A study at the University of Western Ontario compared volunteers who ran four to six 30-second sprints with four minutes' rest (just like Gibala's HIT workout for cycling) with another group running steadily for 30 to 60 minutes at a time. After six weeks of training three times a week, both groups made identical gains in endurance and lost similar amounts of fat. In the "long runs" group, the endurance gains came from increases in the amount of blood pumped by the heart; in the HIT group, almost all the gains came from the muscles themselves, which improved their ability to extract oxygen from circulating blood. Since it's important to have a healthy heart *and* healthy muscles, this suggests you shouldn't rely exclusively on HIT workouts. As with cardio and weights, a mixture is best.

High-intensity exercise is generally thought to carry some risks, so sedentary or older people should check with a doctor before trying HIT. Interestingly, though, University of British

WORKOUTS

The guiding principle of HIT is that the shorter the workout, the higher the intensity you need to reap the benefits. "Basically," McMaster University's Martin Gibala says, "you need to get out of your comfort zone." Start by trying a HIT workout once or twice a week.

The Street-lighter: For a sedentary person who gets winded walking around the block, HIT can be as simple as walking more quickly than usual between two light poles. Then back off, and repeat after you have recovered.

The Classic: Go hard for one minute, then recover (either by slowing down or stopping completely) for one to two minutes. Repeat 10 times. This is a staple workout for a wide range of abilities, suitable for any cardio activity.

The Timesaver: Gibala's protocol of 30 seconds of all-out cycling four to six times with four minutes rest is the shortest workout shown to be effective. But achieving the necessary intensity outside the lab is extremely challenging, so it's best suited to experts and those capable of extreme self-punishment.

All these workouts should be preceded by a gentle warm-up of at least 5 to 10 minutes.

Columbia researcher Darren Warburton has studied HIT training in cancer and heart disease patients and found that these higher-risk populations can also benefit safely from HIT.

There is one catch—the disclaimer at the end of the infomercial, if you will. To cram the benefits of an hour-long workout into a few short minutes, you also have to compress the effort you would have spent. "That's the trade-off," Gibala says. "Going all out is uncomfortable. It hurts." But at least with this approach, it's over quickly.

Can exercise increase my risk of a heart attack?

Danny Kassap was one of the fittest people in the world when he was felled by a heart attack just several kilometers into the 2008 Berlin marathon. The 25-year-old Canadian marathon star survived thanks to a spectator who immediately began CPR, but a few weeks later, Alexei Cherepanov wasn't so lucky. The 19-year-old hockey prospect collapsed and died during a game in Russia's Continental Hockey League. News like that inevitably makes us wonder whether we're tempting fate each time we lace up our running shoes. In the wake of Cherepanov's death, the Continental Hockey League mandated "in-depth medical examinations" for young players currently in the league and compulsory exams for all players in the league's junior draft. But it's not clear that such preventative measures can make much difference.

There's no doubt that, during exercise, your risk of a potentially fatal "cardiac event" is elevated, says Paul Thompson, a cardiologist at Hartford Hospital in Connecticut and a leading researcher on the topic. However, it's equally well established that any risks you incur during an hour of exercise are dwarfed by the reduced risk of a heart attack—as much as 50 percent lower, according to the American Heart Association—during the other 23 hours of the day.

The problem is that the man-bites-dog nature of rare events, like the death of a young athlete, sticks in the mind. "These events really do have a chilling effect on people's desire to exercise," says University of Toronto epidemiologist Donald Redelmeier. To put the risks into context, Redelmeier and a colleague analyzed marathon results from more than three million runners. They found that about two deaths occur for every one million hours of aerobic exercise—a rate that isn't significantly different from the baseline hourly risk of being alive for the average 45-year-old man. The study, published in the *British Medical Journal* in 2007, also found that when cities close roads for a marathon, the chance that a traffic death will be averted is nearly twice as high as the chance that one of the runners will die.

Of course, statistical arguments aren't much help if it's your life on the line, so it's natural to look for ways of screening out the risk. Indeed, autopsies after such deaths reveal pre-existing heart abnormalities in about 94 percent of cases, Thompson says. But it doesn't necessarily follow that we can screen for these abnormalities, because they're so common in the general population. About 10 percent of healthy athletes display abnormal electrocardiograms—and upon further examination, "you keep finding more little abnormalities," Thompson says, producing an impractically high rate of false positives.

Redelmeier agrees: his analysis indicates that any screening program would need to be exceptionally accurate and inexpensive to be worthwhile. Otherwise the resources would be better spent elsewhere, such as improving paramedic staffing at events like marathons to respond to the all-but-unavoidable cardiac events that will sometimes occur. In Kassap's case, the cause turns out to have been myocarditis—an inflammation of the heart caused by a virus, which no screening could have predicted. "People want a riskless society," Thompson says. "So I tell them to go to bed alone."

Will exercising in cold air freeze my lungs?

The strangest story that Michel Ducharme, a scientist with Defence Research and Development Canada, has encountered is the cross-country skiers who were swallowing Vaseline to coat

DO YOU NEED TO SEE A DOCTOR BEFORE EXERCISING?

The Physical Activity Readiness Questionnaire (PAR-Q) was originally developed by the British Columbia Ministry of Health and is now widely used around the world. If you answer yes to any of the seven questions, you should see a doctor before heading to the gym.

- Has your doctor ever said that you have a heart condition *and* that you should only do physical activity recommended by a doctor?
- Do you feel pain in your chest when you do physical activity?
- In the past month, have you had chest pain when you were not doing physical activity?
- Do you lose your balance because of dizziness or do you ever lose consciousness?
- Do you have a bone or joint problem (for example, back, knee, or hip) that could be made worse by a change in your physical activity?
- Is your doctor currently prescribing drugs (for example, water pills) for your blood pressure or a heart condition?
- Do you know of *any other reason* why you should not do physical activity?

their airways as a protective measure against cold air. "That's just crazy," he says—and it's entirely unnecessary. Ducharme is the researcher whose work led to a major revision of the wind-chill scale in 2003, thanks to the efforts of impressively dedicated volunteers who sat in a frigid wind tunnel until their faces developed the first stages of frostbite. And he firmly dismisses the idea that your lungs will suffer from contact with cold air. "The heat exchange is very quick," he says, "and there's no evidence of any risk of freezing tissue."

This may be cold comfort for people who swear they are overcome by coughing fits or throat pain when they exert themselves in subzero conditions. For a long time, people have blamed "exercise-induced bronchoconstriction" (EIB), an asthma-like narrowing of the airways that leads to shortness of breath and coughing, on cold air. The condition affects between 4 and 20 percent of the population—but it's the dryness of the air, not its temperature, that triggers the response.

The dryness-versus-coldness debate has been raging in scientific circles for many years, but recent experiments by Kenneth Rundell, a researcher at Marywood University in Scranton, Pennsylvania, who spent 10 years as an exercise physiologist with the United States Olympic Committee, have essentially settled it. In a study published in the journal *Chest*, Rundell found that warm, dry air triggered the same response as cold, dry air in 22 EIB sufferers riding stationary bikes. The reason is that the cells that line our airways are highly sensitive to dehydration, and breathing hard during exercise means more dry air rushing past these cells.

There are some makeshift solutions. For example, wearing a scarf or balaclava over the mouth can moisten the air as it is inhaled. "That makes breathing more difficult," Rundell notes, so it's less useful for skiers or runners in competition but may be fine in training. Commercial heat-exchange masks, which accomplish the same thing with less breathing resistance, are also available. If the EIB symptoms are serious—and confirmed by a lung-function test administered by a doctor—asthma medication can provide relief.

For most people, it's safe to assume that, short of an asthma attack, exercising outside in the dead of winter is perfectly safe. Even without EIB, some people do experience a burning sensation in their throat or upper airways when they exercise in the cold, Rundell says, "but that's just a response of the nerve endings." In other words, even if it feels like you're freezing your lungs, you're not—so you might as well keep going.

When is it too hot to exercise?

Heat is more than just an inconvenience, as we're reminded almost every summer when we hear the inevitable reports of a football player dropping dead from heat exhaustion during a grueling practice. Since 1960, there have been 128 heat-related deaths among football players at the high school, college, and pro levels—and there's really no excuse for this to keep happening.

The basic steps to avoid danger are straightforward: stay hydrated, schedule workouts for the coolest parts of the day, stay out of the sun, shorten workouts, take more rest, and reduce intensity. But how hot does it need to be for these steps to become important? That depends on who you are and where you are. If you're obese, out of shape, or dehydrated, you'll be much more susceptible to heat exhaustion. And if you live in a part of the country that's usually cool, you'll be more at risk in a sudden heat wave than people farther south who are acclimated to hot temperatures.

One reassuring fact to note is that we're generally pretty good at automatically adjusting our effort under hot conditions. In fact, when researchers at the University of Cape Town asked volunteers to cycle for 20 kilometers in 95°F (35°C) temperatures, they found that the volunteers' brains adjusted to the hotter-than-usual conditions right from the start, before they'd had a chance to start overheating, by automatically signaling fewer muscle fibers in the legs to contract. "When it's hot, you don't wait 20 minutes to slow down, you slow down within a minute," says Ross Tucker, the lead author of the study.

Still, when the conditions are unusually hot, many people do manage to push beyond their limits. In 2010, the medical director of the Twin Cities Marathon in Minnesota, William Roberts, published an analysis of eight recent marathons where hot conditions caused "mass casualty incidents" that overwhelmed local health facilities and caused unacceptable delays in emergency care. He found that most of these races started with weather conditions deemed acceptable under the guidelines published by the American College of Sports Medicine. The problem is that the races took place in cities like Boston, Chicago, London, Rotterdam, and Rochester, N.Y.—all northern cities where temperatures are relatively cool for all but a few months each year.

As a result, mass casualty incidents occurred when the "wet bulb globe temperature" (a corrected scale that factors in air temperature, humidity, and solar radiation) was a relatively modest 70°F (21°C) at the start of the race. Crucially, these races all took place in the spring or fall, rather than in the middle of summer when runners would have been used to the heat. In other words, context matters: the actual temperature is less important than how well you're prepared for it.

Should I avoid exercising outside when air pollution is high?

No surprise, the air pollution in cities is bad for you. And exercise makes it worse, since you breathe in a greater volume of air and bypass the natural filtering of the nasal passages by inhaling through the mouth. That means that when city officials announce a smog alert, there are compelling reasons to think twice about vigorous outdoor exercise anytime between morning and

DEALING WITH HEAT

Acclimatize: Research shows you significantly improve your heat tolerance after 10 to 14 days of exposure. But it's not enough to just sit on the porch fanning yourself—you literally have to sweat. A 2009 San Diego State University study had eight volunteers exercise for 10 days, 90 minutes a day after having Botox injected into one arm to block sweating. At the end of the study, the sweat glands in the sweaty arm were producing 18 percent more sweat—a sign of good acclimation—while the glands from the arm that remained dry became less productive.

Breathable clothing: A 2010 study in the journal *Applied Ergonomics* put the claims of athletic apparel makers to the test, with volunteers exercising for an hour wearing either a cotton T-shirt or a polyester/elastane blend. As expected, the polyester shirt permitted greater sweating efficiency, while the sweat-soaked cotton shirt weighed 50 percent more. Interestingly, though, the cotton shirt didn't actually make the subjects any hotter—so if you're committed to your retro gear, you can stick with it. Whatever material you choose, avoid overdressing, or at least use layers that you can strip off as you heat up.

Dunk yourself: If you do find yourself overheating, the quickest way to cool down is to plunge into the pool—and you don't necessarily need to subject yourself to an ice bath to cool down quickly. Some researchers now argue that pleasant water temperatures of 75 to 79°F (24 to 26°C) are just as effective as colder temperatures and can bring heat stroke victims back to safe temperatures in less than three minutes. The key: the warmer water doesn't constrict the blood vessels under your skin, allowing more efficient heat transfer.

afternoon rush hours, when the levels of most pollutants tend to be highest. But with some careful choices about when and where you go, you can still get in a good—and safe—workout.

The basic problem is that sucking in a mix of gases and particles irritates your airways, which can result in coughing fits and difficulty breathing. But doctors now recognize that these symptoms represent just one part of a larger problem. "All the blood in the body courses through the lungs to pick up oxygen," says Ken Chapman, director of the Asthma and Airway Centre at Toronto's University Health Network. "So if the lungs get inflamed, those inflammatory signals get carried throughout the body." As a result, emergency room doctors deal with higher numbers of serious problems such as strokes and heart attacks on high-pollution days.

High-traffic areas are the most problematic. Australian researchers recently asked test subjects to jog back and forth alongside a four-lane highway and found elevated blood levels of volatile organic compounds, commonly found in gasoline, after just 20 minutes. But pollution levels drop exponentially as you move away from a roadway, according to a 2006 study in the journal *Inhalation Toxicology.* Even just 200 yards from the road, the level of combustion-related particulates is four times lower, and trees have a further protective effect—so riverside bike trails, for instance, have dramatically lower pollution levels than bike lanes along major arteries.

If you usually commute to work on foot or by bicycle, you might think that you'd be better off driving or taking the bus when pollution levels are high. But it's not clear this is always a good choice (not to mention the fact that choosing to drive makes the pollution worse for everyone). A Danish study in 2001 measured pollution exposure while driving or biking along identical routes in Copenhagen. It turned out that the air inside the cars was bad enough that, even taking into account that cyclists were on the road longer and breathing more deeply, the drivers were worse off. On the other hand, an Irish study in 2007 found that cyclists were worse off than bus riders because of their heavy breathing, even though the air on buses was worse than the air outside.

This all sounds a bit confusing—fortunately, researchers at the University of Utrecht crunched a massive set of data to estimate the overall pros and cons of a large number of people switching from driving to biking. On average, sucking in pollution on a bike would shorten your lifespan by 0.8 to 40 days, and the increased risk of a traffic accident would knock another five to nine days off, they concluded. But the health benefits from improved physical fitness would *extend* lifespan by 3 to 14 months!

When you look at it that way, the trade-off seems worthwhile. But it doesn't have to be an all-or-nothing proposition: changing your route to avoid high-traffic areas, changing your schedule to avoid the middle of the day, or simply going a little easier can all reduce the impact of bad air and still let you get a workout in.

How will exercise affect my immune system?

Although scientists aren't sure exactly how it works, recent studies offer plenty of evidence that regular exercise really does strengthen immune function—a claim that can't be made for most of the pills and potions whose sales spike each year when cold and flu season starts. But like any powerful medicine, exercise carries the risk of an overdose. "It's what experts call the 'J-curve' hypothesis," says Brian Timmons, a researcher at McMaster University's Children's Exercise and Nutrition Centre. "Moderate intensity is good, but too much exercise is not so good."

This effect is illustrated in a 2005 study from researchers at the University of Illinois, in which three groups of mice were infected with a flu virus. One group was sedentary, another exercised for 20 to 30 minutes a day, and an extreme group exercised for 2.5 hours a day. The moderate group had an 82 percent survival rate, almost double the 43 percent rate of the mouse potato group that didn't exercise at all—clear evidence that exercise boosted immune function. The heavy-exercise group, on the other hand, had a reduced survival rate of just 30 percent—lower than both the others. The researchers suggest that physical activity tinkers with the balance between two types of immune cell that either promote or inhibit inflammation. Moderate exercise tilts the balance to limit excess inflammation. But prolonged, intense exercise suppresses inflammation too much, preventing the immune cells from doing their job.

Translating the definition of "moderate" exercise from mice to humans is a challenge. Certainly, those following standard exercise guidelines (for example, half an hour of moderate to vigorous exercise five times a week) are bolstering their defences against infection. In fact, in a study of 2,300 runners training for the Los Angeles Marathon, only those training more than 60 miles per week had an elevated risk of catching a cold. "If I'm swimming in the morning then running at night, or doing a marathon, that's where you're going to see a dip in immune function," Timmons says.

Interestingly, the immune-boosting benefits of exercise seem to show up almost instantly. Another mouse study by researchers at Iowa State University, published in 2009, compared the effects of a 14-week moderate exercise program with the effects of a single 45-minute treadmill run done just 15 minutes before being infected with the flu virus. As expected, the habitual exercisers fared best over the following 10 days, showing the greatest reduction in symptoms and virus load. But surprisingly, the mice that had done just one run still fared significantly better than the sedentary controls, although the benefits faded after a few days. This is worth bearing in mind next time you're headed for a high-infection-risk zone like an airplane: working out the day before or even the morning of the flight could help you ward off bugs.

Human studies have found similar results. For example, a year-long University of South Carolina study of 547 adults found that those who exercised moderately caught 20 percent fewer upper respiratory tract infections, with the biggest benefit at the start of cold and flu season in the fall. Other epidemiological studies have made similar observations. So if you're hoping to stay healthy over the winter, the evidence suggests that you'll get more benefit from hitting the gym than from stocking up on orange juice.

Is motivation to exercise genetic?

When Colorado-based Atlas Sports Genetics started offering its "SportsGene" test in 2008, zealous parents rushed to uncover their children's athletic destiny. "The results have helped us immensely," one Texas mom wrote in a testimonial. "We have changed his extracurricular activities to be more in line with the test results." Atlas offers a $169 test of the ACTN3 gene, which, according to an Australian study from 2003, can indicate whether you're best suited to endurance sports, sprint and power sports, or a mix of the two. Whether or not the test imparts useful information—which is still very much up for debate—you probably shouldn't choose how your two-year-old spends his or her playtime based on dreams of future athletic glory. But it's impossible to deny that your genes do play a role in your athletic destiny.

According to a 2006 study of over 85,000 twins in seven countries, about 62 percent of the variation in exercise participation seems to be inherited. This could be because of personality traits that run in families—people who are self-disciplined tend to exercise more, while those who are anxious or depressed exercise less—or physiological differences such as the production of feel-good dopamine after vigorous exercise. The genetic tendency to lose weight or gain muscle could also make some people more likely to exercise than others.

All of this suggests a certain inevitability: either you were born to exercise or you weren't. But in the years since that study, something interesting has happened: the search for the "exercise gene" has run aground. Several studies have analyzed the DNA of thousands of people, looking for the sequences that predict exercise behavior—and they've found not one but many. A 2009 study of 2,600 Dutch and American adults found 37 different DNA regions that were linked to exercise, and these regions were entirely different from the dozens of regions identified in previous studies. In other words, there isn't an exercise gene—there are hundreds of different genes that combine to influence every aspect of our behavior. Nobody has all the "good" genes, but neither does anybody have all "bad" ones. So even if you have difficulty, for example, with losing weight, you're likely to be ideally suited for some of the other mental and physical benefits of exercise.

The idea that your exercise destiny is preordained took another blow from a 2009 study that examined the links between physical fitness and intelligence in 1.2 million Swedish men who enlisted for military service between 1950 and 1976. Among these men were 6,294 twins, which allowed the researchers to separate the effects of nature from nurture. They found that those who increased their cardiovascular fitness between the ages of 15 and 18, a time when the brain is developing rapidly, scored better on cognitive tests and went on to greater educational achievements later in life. Crucially, more than 80 percent of the differences between subjects were explained by environmental factors, while less than 15 percent could be attributed to genetics—a powerful illustration that, while genes might affect how our bodies respond to exercise, the choice of whether or not to exercise still resides with each of us. So if you've been blaming your DNA when you slack off your exercise routine, you'll have to find a new excuse!

How long does it take to get unfit?

Let's start with the good news: researchers have consistently found that it takes less work to maintain your level of fitness than it did to get there in the first place. That means that when life intervenes—holiday travel, exams, deadlines at work, and so on—you can temporarily scale back your workout regimen without losing your hard-earned fitness. But the clock is ticking: Danish researchers found that after just two to three weeks, subjects who reduced physical activity showed worse insulin sensitivity and a decreased ability to burn fat.

In 2008, Paul Williams of the Lawrence Berkeley National Laboratory in California published surprising results about a phenomenon he called "asymmetric weight gain and loss," based on the experiences of 55,000 runners in his National Runners' Health Study. Put simply, he found that you gain more weight when you stop exercising than you lose when you subsequently resume the identical exercise program. "In other words," he says, "if you stop exercising you don't get to resume where you left off." Falling off the exercise wagon for a few weeks may just add a pound or two, but if it happens every year it can lead to steady accumulation of weight even though you're working out diligently for the other 50 weeks of the year. Williams found that, after a break in exercise, women didn't start losing weight again until they were running at least 10 miles a week, and men had to hit twice that total. Once they exceeded that level, the subjects were able to start reversing weight gained over holiday and other breaks.

Everyone faces unavoidable time crunches now and then, so the useful question is how long you can afford to slack before the effects are noticeable. The literature on "detraining," or loss of fitness, is surprisingly complex, because different adaptations to your muscles, heart, and metabolism fade away at different rates. As a rough rule of thumb, the evidence suggests that you retain endurance gains for about two weeks without training, but by the time you hit four weeks you're back to baseline. The drop-off happens sooner if you've just started exercising, while long-time gym rats have structural adaptations that will endure for several months, like a larger heart and more capillaries to take oxygen to their muscles.

There are some strategies that can help preserve your fitness when you're time-crunched. A series of classic studies in the 1980s showed that you can get away with working out fewer times per week, and with doing shorter workouts—as long as you maintain or even increase the intensity. In fact, subjects who were used to training six times a week were able to maintain key fitness indicators such as heart size and oxygen uptake by doing just two high-intensity workouts a week. This is similar to the strategy used in high-intensity interval training (see p. 12).

Interestingly, taking a break from strength training actually leads to an increase in explosive power for a few weeks, as newly strengthened and rested muscles take advantage of elongated tendons. But don't be fooled: University of Tokyo researchers found that the muscles of volunteers who completed a three-month strength training program shrunk back to pre-training size after a month of rest, even though the neuro-muscular strength gains (see p. 7) persisted for another few months. Just like aerobic exercise, it's much easier to maintain strength than it is to rebuild it from scratch, so try to find time for at least a few short workouts even when you're busy.

CHEAT SHEET: GETTING STARTED

- It takes three months of hard training to see significantly bigger muscles, and six weeks to boost endurance, but health and performance gains on a cellular level start within a few days.
- Aiming for a minimum of 30 minutes of moderate exercise five times a week, in bouts as short as 10 minutes, will boost your health, but more is better.
- Your body can be "set" to build strength or aerobic fitness during any given workout, but not both at once, so start your workout with the exercises you're focusing on that day.
- A few minutes of short, intense intervals can produce the same effect as a long, slow aerobic workout—but the intervals have to be hard.
- About two heart-related deaths occur for every million hours of aerobic exercise, due to pre-existing abnormalities in 94 percent of cases. Not exercising is far more dangerous for your heart.
- Cold air can't freeze your lungs, but dry air can trigger an asthma-like response in between 4 and 20 percent of people. You can moisten incoming air by wearing a scarf or breathing mask.
- Be cautious doing prolonged hard exercise in temperatures above about 70°F (21°C). It takes about 10 to 14 days to acclimatize to hotter conditions.
- You take in more polluted air when you're breathing hard. Exercising in the early morning or evening and staying a few hundred yards away from busy roads can reduce your exposure.
- Athletic ability and desire to exercise are influenced by many different genes; recent studies suggest that more than 80 percent of the differences among us are environmental, not genetic.
- You can retain fitness for about two weeks without training before significant losses occur. A couple of short, hard workouts each week can preserve fitness for longer.

2: Fitness Gear

In 2009, Americans bought an astonishing $72 billion worth of sports and fitness gear—about three times as much as they spent on books. Some of that was apparel (led by running shoes); some was equipment (led by treadmills); and some, inevitably, was late-night infomercial fodder like the Dumbbell Alarm Clock, which for $30 will buzz until you complete 30 biceps curls. All this equipment has the goal of making your workouts more pleasant and more productive—but in a hyper-competitive and poorly regulated market, it's a challenge to figure out which miraculous claims are backed by solid evidence, and which are nothing but hype.

Is running on a treadmill better or worse than running outside?

Running on a treadmill is a bit like making bread with a bread machine: purists say you're missing the essence of the experience, while pragmatists say the convenience and results speak for themselves. Even critics have to admit that running indoors on a treadmill offers plenty of benefits, like controlled temperature, good footing, even pacing, and maybe even a big-screen TV nearby. But that doesn't tell us whether you're working the same muscles on the treadmill as you would outside—a crucial question if, for example, you're using the treadmill over the winter to prepare for a 10K road race in the spring.

"This is actually a very ugly question, to which there's no definitive answer," says Colin Dombroski, a pedorthist (someone who specializes in footwear for athletes and people with foot or lower limb problems) at the Fowler Kennedy Sport Medicine Clinic in London, Ontario. More than 30 years of research have produced two schools of thought. One holds that treadmill running is fundamentally different because you're just whipping your feet from back to front without moving your center of mass. The other maintains that, as long as the treadmill is moving at a constant speed, there's no physical difference other than the lack of wind resistance.

The most recent attempt to unravel this mystery comes from researchers at the University of Virginia, who used high-speed cameras and special treadmills with force-measuring surfaces to compare joint motion and impacts. The results, published in the journal *Medicine & Science in Sports & Exercise* in 2008, show that there are some statistically significant differences, such as knee orientation and peak force. But overall, the researchers concluded that the biomechanics of running on a treadmill are close enough to "overground" running that the differences don't matter.

This is the same conclusion that many running coaches have reached through first-hand experience. "There are differences, but they're very minor," says Peter Pimm, a Toronto distance-running coach who has guided Olympians as well as recreational runners for more than 25 years. To compensate for the lack of wind resistance, Pimm has his runners set their treadmills at a 1 percent incline.

One key advantage of treadmills, he says, is that they're generally softer than sidewalks and roads. "People often notice that it's gentler on the knees," he says. "So if you've had joint injuries, it may even allow you to train more." This may also be the treadmill's greatest weakness: if you stick to the treadmill all winter, the lack of pounding means you don't build the same level of specific muscular endurance that you would from running outdoors. So if you jump straight into a road race, Pimm says, "you may develop muscle soreness and tightness earlier in the race than you otherwise would." The same caveat applies to running on grass and trails, Dombroski says. The treadmill won't develop the stabilizer muscles that keep you upright on uneven surfaces.

In any case, the answer is simple: variety. Try not to rely exclusively on the treadmill—and if weather does keep you

SETTING THE TREADMILL INCLINE

With no air resistance to slow you down, running on a treadmill is slightly easier than running outside on a calm day. Researchers at the University of Brighton found that running at a 1 percent incline consumes about the same energy as outdoor running on a flat road for speeds around seven minutes per mile. At speeds of nine minutes per mile or slower, a half-percent incline is sufficient.

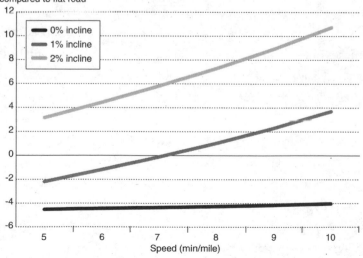

Extra energy needed
compared to flat road

- 0% incline
- 1% incline
- 2% incline

Speed (min/mile)

indoors for several months, make sure you start with a few easy outdoor runs before jumping into a 10K race.

Is the elliptical machine just as good as running?

More than 23 million Americans used elliptical trainers in 2007, according to the Sporting Goods Manufacturers Association, triple the number that used them in 2000. Canadians have also eagerly adopted the low-impact, pseudo-jogging machine. But researchers are still debating how the elliptical stacks up to its higher-impact rivals—whether it really does reduce injury risk and how hard a workout it provides.

A pair of representative studies, both from 2005, illustrate the debate. In one, researchers from the University of Scranton asked volunteers to exercise at a level that felt "hard" on either a treadmill or an elliptical machine. The subjects had higher heart rates and burned more energy on the treadmill, suggesting that they unconsciously chose to push harder when running. In contrast, when researchers at the University of Idaho had volunteers exercise at identical heart rate and energy consumption levels on an elliptical, a treadmill, and a recumbent bike, the volunteers reported that the elliptical felt easiest. Subsequent studies haven't managed to settle the debate—and that suggests that any actual differences are probably too small to matter.

An Irish study published in the *Journal of Sports Medicine and Physical Fitness* is the only one to tackle the practical outcomes that most people are concerned about: fitness and weight loss. A group of 24 women performed a 12-week fitness program using an elliptical, a treadmill, or a stair-climbing machine. Cardiovascular fitness increased and body fat percentage decreased to the same degree in all three groups, confirming that any calorie-burning differences between these machines aren't significant.

There are real differences, though. A recent study by Western Washington University exercise scientist Kathleen Knutzen used elliptical machines with force plates on the pedals to determine that, even at a fast stride, the forces on the lower legs were comparable to walking, and two or three times smaller than running. "That's a real benefit if you're prone to repetitive stress injuries," Knutzen says. But the constrained elliptical stride calls on muscles in a different way than running freely would, much like the differences between lifting free weights and using a weight machine. That means that the elliptical will never be a perfect substitute for running—although it's an ideal cross-training activity for injured runners whose overuse injuries benefit from a slightly different set of motions.

The elliptical also offers an opportunity to get your upper body involved by pumping the arm handles, though studies suggest that most people use the arm handles mostly for balance. Constance Mier and her colleagues at Barry University in Florida have found that using the arm handles generally makes only a small difference to calorie burning and perceived effort, and no difference to heart rate. Still, the two actions are sufficiently different that alternating between them might allow you to last longer—and thus get a better workout—when using the elliptical.

Ultimately, the elliptical is a perfectly good way to get fit. But there's also something to be said for using your workout to develop motor patterns that you'll use in the rest of your life. So unless your goal is to master the art of moving your feet in little ellipses, it makes sense to include some more functional activities like walking, running, or biking. "Mix it up," Knutzen advises. "Don't do the same thing every day."

Do I really need specialized shoes for walking, running, tennis, basketball, and so on?

Kids everywhere have grown up believing Spike Lee's famous pitch for Nike's Air Jordans: "It's gotta be the shoes!" Over the past few decades, shoe companies have poured millions of dollars into researching how shoes affect the motion of your feet and the forces transmitted to your joints, and into developing high-tech materials to make shoes more comfortable, durable, and functional. There's no doubt that sports shoes have come a long way as a result, but how much is just hype?

For many sports, the most obvious benefit that specialized shoes provide is traction. In sports like soccer that are played on grass, cleats provide a major advantage for starting, stopping, and changing direction. Traction also has benefits in less expected contexts, like the weight room. A 2007 study found that test subjects were able to lift more weight and spend less energy doing it in shoes with good grip compared to shoes with more slippery soles.

Court sports like tennis and basketball also place very specific demands on the foot. A 2008 study in the *American Journal of Sports Medicine* found that jumping and sudden changes of direction put twice as much pressure on the heel as running in a straight line. Those forces are reflected in the structural differences between court shoes, which are optimized for lateral motion, and running shoes, which are designed to roll from heel to toe with each step. That's why running shoes are neither comfortable nor effective for playing court sports—and it means that, if you really want to own just one pair of athletic shoes, you'll have to make some decisions about which activities are most important to you.

But what about injury protection? The research is much less clear here. Even designs that seem to have obvious benefits—like the high-top basketball shoe as a way to reduce ankle injuries—remain unproven. There's plenty of evidence that high-tops make your ankle feel more stable and restrict its maximum range of motion. But the few studies that have actually tried to prove a link between basketball shoe type and injury rate have been inconclusive, according to a 2008 review conducted by the Cochrane Collaboration, an independent, nonprofit group that assesses evidence-based health care research.

Which brings us to the most controversial topic of all: running shoes. Just like high-top basketball shoes, there's plenty of research into how different types of running shoe alter the forces transmitted through your legs each time your foot smacks the ground. But it has been very difficult to take the next step and show that certain types of running shoe decrease (or increase, for that matter) injury rates.

Researchers at the Allan McGavin Sports Medicine Centre in Vancouver have used training clinics for the Vancouver Sun Run 10K, an annual race that attracts more than 60,000 runners, as a laboratory to study the training of thousands of recreational runners over the past two decades. Injuries still occur, says lead author Jack Taunton, who was the chief medical officer of the 2010 Olympics, "but as we've seen shoes get better over the years, they're less often found to be the main factor." One correlation that Taunton found is that men whose shoes were more than four months old were more likely to get injured. For women, the effect was noticeable only once their shoes were more than six months old, presumably because men are generally heavier and thus compress the shoe cushioning more quickly.

On the other hand, a 2010 study from Taunton's laboratory found that prescribing certain types of shoes ("neutral," "stability," or "motion control") to runners based on their foot posture didn't decrease injury rates in the subsequent 13 weeks. If anything, the runners assigned to heavy "motion control" shoes experienced more injuries than the control group who received randomly assigned shoes.

What does seem increasingly clear is that the rival running-shoe technologies touted by manufacturers don't make much difference. A study by researchers at the University of Texas at El Paso, published in 2009 in the *British Journal of Sports Medicine*, found no difference in the effects of air, gel, and spring-type cushioning on the performance of running shoes worn for 200 miles or more. Still, if you head out for a run, you're much better off in a pair of running shoes—whatever the type—than you would be in a pair of tennis or basketball shoes. Unless, that is, you've decided to go barefoot . . .

Will running barefoot help me avoid injuries?

We were born to run—barefoot. That's the buzz these days, spurred in part by Christopher McDougall's 2009 bestseller, *Born to Run,* and by Harvard anthropologist Daniel Lieberman's argument that distance running shaped the evolution of the human body. Now biomechanical studies are beginning to emerge that—by some accounts—finally offer scientific proof that you'd be better off ditching your footwear or running in "minimalist" shoes that do little more than protect your soles. But these studies need to be interpreted with caution.

The question of whether modern running shoes increase or decrease injuries remains shrouded in uncertainty. In a 2009 *British Journal of Sports Medicine* article, Australian shoe researcher Craig Richards argued that no study has ever succeeded in showing that modern running shoes reduce injuries.

"Shoe researchers and manufacturers will try and bamboozle you with the results of hundreds of biomechanical studies," he says. But these studies simply show that shoes change the forces acting on your feet—and no one knows for sure how those forces relate to injury rates.

In 2010, Lieberman published in the journal *Nature* the results of a study of American and Kenyan runners. His two main conclusions were, first, that barefoot runners are more likely to land on the middle or front of their feet, while shod runners generally land on their heels; second, that the peak vertical force upon impact is three times greater in shoes than it is barefoot. Newspaper reporters took this as proof that barefoot running is "better"—but in fact, this is precisely the kind of biomechanical study that Richards had dismissed as useless, because it says nothing about injury rates.

Another study, this one from Casey Kerrigan and her colleagues at the University of Virginia, made even bigger headlines in early 2010 with the claim that running shoes are worse for your feet than high heels. What this study actually found was that running shoes create greater torque on the hip and knee joints than barefoot running. They then compared these results to an earlier study of walking in high-heeled shoes and made a very dubious comparison between the two.

There's no doubt that thinking on footwear has evolved in the last decade. For instance, plush cushioning is no longer considered the ultimate defense against injury. "I wish running companies would stop rattling on about 'gel' and 'air' and so on," says Simon Bartold, an Australian shoe researcher who consults for Asics, a sports shoe manufacturer. Newer shoes reflect this thinking, he says: Nike has introduced the bare-bones Free, for example, and Asics has quietly abandoned the concept of "motion control." But there's no evidence that runners of all shapes and sizes would benefit from simply giving up on shoes. Of course, Bartold works for Asics, so we can't trust him, right? Well, Lieberman's study was funded by Vibram, which makes the barefoot-simulating FiveFinger. And Kerrigan and Richards have both founded their own minimalist shoe companies.

The point isn't that the new barefoot studies are bad. On the contrary, they offer valuable information about how we run, and they do indeed suggest that injury-prone runners might benefit from some cautious experimentation with barefoot (or nearly barefoot) running. That might involve jogging barefoot on grass for five minutes, a couple of times a week, and building up from there. But if we're holding the barefoot advocates to the same standard of "proof" that we demand from the shoe companies, let's be clear: any link between barefoot running and reduced running injuries remains unproven for now. (The negative effects of stepping on something sharp, on the other hand, are extremely well established!)

Will compression clothing help me exercise?

From the sleeve on Allen Iverson's shooting arm to marathoners racing in knee-high socks, there's a lot of tight clothing in the upper echelons of sport these days. But compression garments have come a long way from the wave of colorful spandex that engulfed gyms in the 1980s. The first wave of tight clothing offered benefits like cooling, sweat management, reduced chafing, and (ahem) better support. Modern compression garments, in contrast, are designed to squeeze your arms and legs hard enough to affect blood flow and stabilize vibration in your muscles, resulting in bolder claims of enhanced power, better endurance, and faster recovery.

These new garments are descended from medical leggings that have been used for decades to treat blood clots and circulatory disorders. The key is that they deploy "graduated" compression: they squeeze more and more tightly as they move farther from the heart, which reduces blood pooling in the legs and speeds the return of blood to the heart. One of the clearest benefits compression offers for athletes is quicker recovery from "delayed onset muscle soreness"—the aches that appear after an intense bout of unaccustomed exercise. Wearing a compression sleeve around the affected muscles seems to help control swelling and, through enhanced circulation, hasten the removal of cellular waste products.

More controversial is the research into explosive movements such as sprinting and jumping. A 1996 paper by University of Connecticut researcher William Kraemer, then at Penn State, found that volleyball players who wore compression socks were able to produce more power in their vertical jumps. One theory is that the physical support offered by the garment reduces unwanted oscillation and jarring of the muscle—that's the rationale behind compression shorts for basketball players, though more recent studies have produced conflicting results.

For runners and cyclists, the key to potential endurance gains is the "calf muscle pump": with each step or pedal-stroke, the clenching and unclenching of the calf muscle squeezes blood back toward the heart. Compression socks covering the calf provide an extra squeeze that enhances this pumping action, speeding the flow of much-needed oxygen to working muscles. The enhanced pump is so powerful, according to Central Queensland University researcher Aaron Scanlan, that some researchers believe there's no additional benefit to wearing full-length leggings rather than just knee-length socks.

Tests of this supposed endurance boost have not yet been conclusive—in part, Scanlan says, because it's so hard to apply exactly the same level of compression to people whose calves have different sizes and shapes. "No one has really figured out the definitive pressure to improve performance," he says. Some studies of runners and cyclists have found improved performance in time trials or changes in physiological measures like muscle oxygenation when the subjects wear compression socks; others have failed to reproduce those results.

Most telling, perhaps, is a study presented by researchers from Indiana University at the 2010 annual meeting of the American College of Sports Medicine. A group of 16 volunteers ran at three different speeds with and without compression socks, and the researchers found no differences in average running economy or in the runners' stride patterns. But when they looked more closely at the individual results, they found that four of the subjects had experienced a significant increase in economy, while four others experienced a significant decrease. Interestingly, the subjects whose economy improved were the ones who had stated in a pre-experiment questionnaire that they *expected* the socks to help them. "There may be a psychological component to compression's effects," lead researcher Abigail Laymon speculates. "Maybe if you have this positive feeling about it and you like them, then it may work for you. It is a very individual response."

Does walking with poles give me a better workout?
It used to be only out-of-season cross-country skiers who walked with poles. Then hikers carrying heavy loads over rough terrain made the transition from hefty walking sticks in one hand to lightweight poles in both hands. Now walking with poles is so

THE CALF MUSCLE PUMP

With each step, the calf muscle squeezes blood back toward the heart, while one-way valves above and below the calf keep it flowing in the right direction. Compression socks add more squeeze to make the pump even stronger, which is why many athletes wear them on long flights to prevent swelling caused by blood pooling in the lower legs.

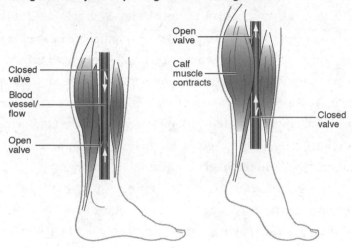

mainstream that it has a new name—"Nordic walking"—and since 1997 the activity has been promoted by the International Nordic Walking Association.

So what do the poles actually do for you? A 2010 study by Italian researchers confirmed what several earlier studies had found: walking with poles can burn about 20 percent more energy without feeling any harder than walking without poles at the same pace. The vigorous arm swing recruits back, shoulder, and other upper-body muscles, burning extra energy. Interestingly, though, all this upper-body action isn't actually propelling you forward. This was shown in a clever study by researchers at the German Sport University in Cologne, who deployed walkers with force-sensing poles on concrete, grass, and a rubberized track. As expected, the walkers had to work harder on the soft grass than they did on concrete—but force generated by the poles remained the same, leading the researchers to conclude that "the effects of poling action on overall propulsion is only marginal." In other words, it's still the legs doing all the important work.

That conclusion holds true only on level ground. The Italian study is one of several to find that pole walking feels easier than pole-less walking at the same pace when going up a hill with a 5 percent incline. Even though pushing on poles doesn't do much to propel you forward on flat land, it does seem to help push you upward on hills, along with improving stability—not a surprising finding, given the longstanding popularity of walking poles among hikers in mountainous areas.

There's an important caveat, though. You don't get the benefits of Nordic walking simply by grabbing a couple of poles and dragging them along behind you. You need to use proper form and put in a good upper-body effort. According to the International Nordic Walking Association, that means leaning forward slightly, rotating the chest with each stride, and following through as you push with the pole so that your arm extends behind you. The poles should be angled at the rear, not held vertically, so that you can push off to work your arm muscles.

Are sports video games real workouts?

This is a question that has launched dozens of studies since the release of Nintendo's Wii gaming system in late 2006, with the results now being debated in lofty academic journals. Most researchers agree that "exer-gaming" does burn significantly more calories than playing traditional video games, but that's really not saying much. The real question is whether they burn enough calories to improve health and fitness outcomes—and whether encouraging kids to play these games inspires them to get outside and try the real thing, or simply keeps them hooked on video games.

In some respects, video-game sports are doing a good job of simulating the real thing—injuries, for example. The *New England Journal of Medicine* reported the first case of "Wiiitis" in 2007, a 29-year-old suffering from tendinitis in his shoulder after playing Wii tennis. Other injuries reported in the medical literature include head trauma in an 8-year-old girl whose brother accidentally hit her while swinging his controller, and, in 2010, the first "Wii fracture" in the foot of a 14-year-old who fell off her Wii Fit balance board.

But what about improving fitness levels? To compare the energy expenditure of different tasks, researchers use "metabolic equivalents of task" (METS). Sitting quietly on the sofa requires 1 MET, while playing tennis for the same period of time typically requires about eight times as much energy, or 8 METS. A 2010 study of 51 students from the University of Waterloo, Ontario, which appeared in the *American Journal of Health Behavior*, found that a traditional video-game version of tennis—Mario Power Tennis for the Nintendo GameCube—required 1.2 METS, barely more than doing nothing at all. Wii Sports tennis, on the other hand, burned a respectable 5.4 METS on average. These results are fairly consistent with earlier studies, given that—just like real sports—the energy you burn depends on how vigorously you play. For example, a small 2007 study of 11 subjects found a more modest expenditure of 2.5 METS for Wii tennis. Another study, published in *Obesity* in 2009, found that active video games like Dance Dance Revolution and even Wii bowling are comparable to moderate-intensity walking.

So far, only three short-term studies lasting between 6 and 12 weeks have tried to address the crucial question of whether active video games can actually lead to improved health in children. According to a 2009 review in the journal *Pediatrics*, none of these found any significant effects on outcomes such as body-mass index (BMI), though such changes would generally take longer than the scope of the studies to show up anyway.

On the other hand, three-quarters of young North Americans spend more than 10 hours a week sitting in front of various screens, and studies have found that they're very unwilling to relinquish that time. With that in mind, even activities that mimic a casual stroll are better than nothing. Scott Leatherdale, the lead author of the Waterloo study, calculates that males who play an hour a day of active video games would burn an extra 483 calories per week—the equivalent of 7.2 pounds of fat per year. "The basic message is that if kids are going to play video games, parents should at least try to get their kids playing games that involve being physically active," he says. "That being said, video games should not replace actual physical activity."

What should I do with wobble boards and exercise balls?

No gym is complete these days without an assortment of oddly shaped and surprisingly expensive balance-training gadgets. Unlike many fads, this one has its roots in solid medical research: wobble boards earned their stripes decades ago in the rehabilitation of ankle sprains. Balance training (including techniques like yoga that don't require any specialized equipment) is now touted for injury prevention and the strengthening of countless small stabilizer muscles that help keep us upright and maintain posture. For those who play court sports involving lots of running and rapid changes of direction, the benefits are clear; for the rest of us, the verdict is mixed.

Two recent studies by University of Calgary physiotherapist Carolyn Emery and her colleagues followed more than 1,000 high school basketball players and physical education students in randomized trials of wobble-board balance training. In both cases, balance training reduced injuries. The rate of ankle injuries among basketball players, for instance, was 36 percent lower in the balance-training group.

Still, critics point to a pair of studies that associated balance training with increased injury risk. The first was a 2000 study of 221 Swedish soccer players that observed four knee ligament injuries in the balance-training group versus just one in the control group—numbers that are too small to draw firm conclusions from. The second was a 2004 study of Dutch and Norwegian volleyball players that saw an increase in knee injuries for players with a prior history of knee problems. "These results are somewhat surprising, and we do need to be mindful of them," says Con Hrysomallis, a researcher at Victoria University in Australia who reviewed 21 similar studies across a variety of sports in a 2007 article in the journal *Sports Medicine.* The two negative studies were the only ones that had subjects throwing and catching balls while on a wobble board, as opposed to simpler exercises, he notes.

The overall message from the studies Hrysomallis reviewed is positive, he says. Studies have consistently found that balance training, along with other "neuromuscular" exercise such as jumping and agility drills, can reduce injury risk in sports such as soccer, basketball, and volleyball. There's also some evidence that simple balancing drills can reduce the risk of falls in older people.

That doesn't mean you should start doing all your exercises on unstable surfaces. If you do a bench press while lying on an exercise ball, you won't be able to lift as much weight, so you'll gain less strength. As a 2007 study of soccer players at the University of Connecticut concluded, too much training on unstable surfaces "may create a hesitant athlete for whom stability is gained at the expense of mobility and force production." Interestingly, another recent study in the *Journal of Strength and Conditioning Research* found that for subjects who already have a lot of experience training with free weights, moderately unstable platforms (such as Bosu balls and Dyna Discs) don't appear to stimulate any additional muscle activation compared with stable surfaces. The authors conclude from this that simply using free weights is enough to keep your "stabilizer" muscles in good shape.

Add up the evidence, and you're left with a familiar message: moderation. Balance training can be a useful tool to help prevent lower-leg injuries, and perhaps to ward off falls as you get older. But if you spend too much time on the exercise ball, you'll be missing out on other training benefits.

Can a mouthpiece make me stronger, faster, and more flexible?

One guy who wouldn't have needed any help from a mouthpiece is Michael Jordan. "When he went up for a dunk, he'd stick his tongue out," says Anil Makkar, the Nova Scotia dentist whose Makkar Pure Power Mouthpiece has become a must-have accessory for professional athletes in a variety of sports. Sticking your tongue out, it turns out, lowers your jaw and brings it forward into a relaxed position. Most of us, in contrast, clench our teeth with effort—and that's why we'd benefit from a mouthpiece that keeps the jaw and facial muscles relaxed, Makkar says.

ADDING BALANCE TRAINING TO YOUR WORKOUT

Balance test: If you can't stand on one leg with your eyes closed for at least 15 seconds, you have poor balance.

STATIC EXERCISES
Stand on one leg for as long as you can. Close your eyes to add difficulty. Progress to one-legged calf raises and squats.

DYNAMIC EXERCISES
To simulate real-life movements, try stepping, lunging, or jumping onto one foot, then balancing.

UNSTABLE SURFACE
Once you're comfortable on a stable floor, try the same exercises on unstable surfaces.

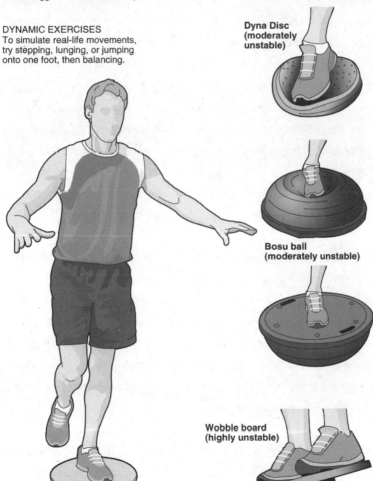

Dyna Disc (moderately unstable)

Bosu ball (moderately unstable)

Wobble board (highly unstable)

Star athletes ranging from Shaquille O'Neal to Terrell Owens have publicly testified to the performance-boosting powers of Dr. Makkar's $2,000 mouthpiece (entry-level models cost $600). Baltimore-based Under Armour's rival mouthpiece boasts a client list that includes several dozen 2010 Winter Olympians and almost 100 National Hockey League players, including Alexander Ovechkin. Before you invest, though, it's worth looking into the research behind the bold claims—because not all "scientific proof" is created equal.

The idea that the position of your jaw can affect the rest of your body stretches back at least to ancient Greek athletes and Roman warriors biting down on leather straps; wounded U.S. Civil War soldiers "bit the bullet" to deal with pain. These days, top sprinters strive to relax their face—as you can see from their jiggling cheeks in slow-motion replays.

Nobody really knows why this should work. There are many theories, including that clenching may stimulate excessive production of the stress hormone cortisol, constrict airways, or interfere with nerve signals traveling from the brain to the rest of the body. Whatever the mechanism, Makkar Athletics reports immediate improvements in posture, flexibility, balance, and strength, and notes that its users report increased endurance and faster recovery. Under Armour makes similar claims and adds faster reaction time.

To back up these startling claims, Makkar funded an independent study at Rutgers University. In a double-blind study, researcher Shawn Arent tested 22 collegiate and professional athletes, all from contact sports where mouthpieces are already used to protect teeth. Each subject was fitted for a standard mouthguard and for one optimized with Makkar's hour-long proprietary technique, and neither the athletes nor the researchers knew which one they were wearing. Arent—to his surprise, he admits—observed small but statistically significant improvements in vertical jump, in peak power produced in a 30-second cycling test, and in the average and peak powers produced during a sequence of 10-second bursts of cycling. The only test that didn't show a significant change was the number of body-weight bench presses.

The research that Under Armour has made available so far is less convincing. For example, a series of 2008 studies conducted at the Citadel, a military college in South Carolina, compares subjects with mouthpieces to subjects with nothing in their mouths, taking no account of the placebo effect. And what's described as "a definite trend for lower cortisol" turns out to mean that cortisol levels were lower in only 11 of the 21 cyclists in the study—barely more than half. A follow-up study of runners in 2009 also failed to find any statistically significant change in cortisol.

All these results are too preliminary to draw any firm conclusions about what different mouthpieces can and can't do. Until more research is completed, anyone investing in these devices is making a leap of faith, not a scientific judgment. But there's enough to suggest that the link between jaw position and physical performance isn't just fantasy. "My sense is that it's real and it could be important, for some sports more than others," Arent says.

Is there any benefit to strengthening my breathing muscles?

It makes perfect sense: to avoid getting out of breath, you should improve your breathing muscles. When you inhale, you use the muscles of the chest wall and diaphragm to suck in air; you then relax those muscles to push air back out. After strenuous exercise, those muscles can fatigue. That's the observation that spurred British researcher Alison McConnell of Brunel University in London to develop the Powerbreathe "inspiratory muscle trainer" in the 1990s, a portable device that looks like an oversized asthma inhaler, designed to strengthen the muscles you use to inhale, just as weights strengthen your arm muscles. Taking 30 breaths through the machine twice a day, and gradually increasing the resistance, is supposed to strengthen the muscles, increase endurance, and make you feel less out of breath.

There's just one problem. Initial studies by a variety of researchers failed to find any benefits from inspiratory muscle training, though subjects often reported feeling less out of breath. After testing classic laboratory measures of aerobic endurance like VO_2max (see p. 63) with no success, many researchers concluded that breathing isn't a limiting factor in endurance after all—that the limits are instead determined by your ability to circulate oxygen through the blood, or your muscles' ability to make use of that oxygen. The exception is people with conditions like chronic obstructive pulmonary disease, where this kind of respiratory exercise has long been standard.

But athletes are more concerned with how they perform in competition than in the lab, and several more recent studies have demonstrated small but significant benefits in sports like swimming, cycling, and rowing. For example, a 2010 study at Auckland University of Technology in New Zealand put 16 competitive swimmers through a six-week Powerbreathe training program; half of them did the real thing, while the other half used the same device but were given a training program that didn't actually strengthen their inspiratory muscles. The experimental group improved by 1.7 percent in the 100-meter freestyle and 1.5 percent in the 200 meters compared to the sham training group, but didn't show significant improvement in the 400 meters.

Swimmers are good candidates for this type of training for several reasons: they have to control their breathing rate to match their stroke, and they have to overcome water pressure to expand their chests when they inhale. Rowers also face the challenge of synchronizing their breathing to their stroke rate. But even cyclists, who don't face the same constraints, have shown improvements of 2.7 to 4.6 percent in time trials ranging from 20 to 40 kilometers after inspiratory muscle training. And a 2010 study by researchers in Hong Kong saw improvement in a series of repeated 20-meter shuttle sprints when the subjects did twice-daily inspiratory muscle training. The runners also warmed up before workouts and testing sessions with a slightly easier breathing routine.

There are several plausible theories of why this form of training should work, the simplest being that stronger breathing muscles allow you to pump in more oxygen when you're tired. Even if breathing rate isn't a limiting factor, it may be that stronger breathing muscles require less oxygen-rich blood to fuel them, allowing it to be diverted to arm and leg muscles. A more subtle possibility is that the primary benefit is how you feel—which would explain the initial failure to find improvements in laboratory studies. If theories about fatigue originating in the brain are correct (see p. 57), then feeling less out of breath could allow you to keep pushing for longer, even if there's no actual difference in the functioning of your lungs or other parts of your body.

The downside of the "it's how you feel" theory is that the performance-enhancing effects may then fade away as you become accustomed to inspiratory muscle training. Clearly, much more research will be needed before we really understand if and how this technique works. Until then, unless you're an elite athlete searching for an extra hundredth of a second, you're probably better off focusing your time and energy on a more traditional form of breathing training with proven results: swimming or running or biking until you're out of breath.

CHEAT SHEET: FITNESS GEAR

- Your stride on the treadmill is the same as it is outside, but you may need time to readjust to harder outdoor surfaces, so do a few outdoor runs before any races. Set the treadmill incline at 0.5 to 1 percent to compensate for the lack of wind resistance.
- Elliptical machines offer a low-impact aerobic workout that is equivalent to running or biking, but they don't develop "functional" muscle patterns. You can prolong the workout by using the arm levers.
- Athletic shoes are optimized for the different movement patterns and playing surfaces in different sports. This boosts performance, but the evidence that the right shoes reduce injuries remains weak.
- Running in bare feet produces a different stride pattern and different stresses on your feet and legs, but there's no evidence yet to link it to a reduction in injuries.
- There's increasing evidence that compression socks and sleeves can help speed muscle recovery after intense workouts. Claims that they boost power and endurance remain unconvincing.
- Walking poles help you burn 20 percent more energy by involving your arms and propelling you up hills—as long as you use proper form and vigorous push-offs.
- "Active video games" burn more energy than traditional games but are generally equivalent only to a leisurely walk.
- Balance training is vital for avoiding the recurrence of ankle and knee problems and may help prevent them in the first place. But you should still do some training on solid ground to maximize strength gains.
- Initial studies suggest that specially fitted mouthpieces may boost performance by a few percent by keeping your jaw relaxed, but the evidence remains patchy.
- "Inspiratory muscle training" to strengthen your breathing muscles appears to boost endurance by a few percentage points, but it's unclear whether the benefits are lasting.

3: The Physiology of Exercise

SIR ROGER BANNISTER, THE EMINENT neurologist whose sub–four-minute mile back in 1954 remains one of the most celebrated examples of human boundary-breaking, put it best: "The human body is centuries ahead of the physiologist." Today's scientists have an exciting array of new tools and techniques that allow them to peer inside the body with more precision than ever before, monitoring how slight changes in training or nutrition affect health and performance on a cellular level. But more often than not, their findings merely confirm the wisdom that coaches and athletes have arrived at through trial and error, rather than producing radical new approaches to getting fit.

Still, it can be helpful to understand how your body responds to exercise and what the various physical sensations it produces mean. One of the most exciting fields of research these days is on the origins of fatigue: old villains like lactic acid are getting left behind, while the role of the mind in regulating the limits of endurance is increasingly studied. Meanwhile, understanding the causes of common nuisances like post-exercise soreness, muscle cramps, and stitches can give you the knowledge you need to avoid them in the future.

What role does my brain play in fatigue?

Imagine crossing the finish line of a 10K running race—or a bike ride or any other activity that pushes you to your limits. You're out of breath, and your heart is thumping. Your legs are burning, you're overheating and dripping sweat, and you feel as though your fuel gauge is on empty. All these factors contribute to your sense of fatigue, but which was the one that actually prevented you from going faster or farther? Scientists have been pursuing the answer to this question for the last century. But according to a radical theory that has been gaining momentum in the last few years, there is no answer—because it's the wrong question.

Researchers test the limits of endurance by putting athletes on a treadmill and gradually increasing the speed until they're forced to stop (or fall off the back of the treadmill). But compare this to what happens in real-life athletic contests. While running a race, you never reach a point where you simply keel over (unless something goes badly wrong). Instead, you're constantly adjusting your effort with the goal of running as fast as you can while ensuring that you complete the distance. So whatever "failure" causes you to fall off the treadmill at the end of a maximal test can't be the same thing that prevents you from running faster over 10K.

What's been missing here is the role of the brain. Instead of our limits being dictated by "peripheral" fatigue—a failure somewhere in the muscles of your legs, the beating of your heart, or the pumping of your lungs—South African researcher Tim Noakes has proposed that a "central governor" in the brain regulates our physical exertions. This governor integrates physiological information from throughout the body—core temperature, blood oxygenation, muscle signals, and so on—along with other data based on previous experience and knowledge of how long you expect to continue. Operating beyond conscious control, it regulates how much muscle you're able to activate, with the goal of holding you back before you reach a state that could damage your heart or other organs.

This doesn't mean that fatigue is imaginary. Your body really does have physical limits—but, if the central governor theory is correct, your brain rarely permits your body to actually reach them. The simplest example of this phenomenon is the finishing sprint that is a nearly universal phenomenon across endurance sports, from novices to world-record holders. No matter how hard you thought you were going, you suddenly find as you approach the finish that your legs can move faster after all. Nothing has changed physiologically—but your central governor allows you to speed up now that the finish line is in sight.

In contrast, if you put subjects in a hot room and ask them to pedal an exercise bike as hard as they can, their power output will be lower than in cool conditions—right from the first pedal stroke. The slowdown happens long before any of the physical effects of heat could be relevant—further evidence that the brain is quietly enforcing a safe "maximal" effort.

This debate between peripheral and central models of fatigue is perhaps the most controversial topic in current exercise physiology. No definitive conclusions are in sight, but there's broad recognition that the brain plays a larger role than previously acknowledged. This role is unconscious, so you can't simply "decide" to push through to your true physical limits—which is probably a good thing. What you can do, though, is gradually teach your brain what your body is capable of. For example, training at your goal race pace not only increases fitness, but also allows your mind to become familiar with the accompanying physiological feedback. You can't turn your central governor off—but with patience you can adjust its settings.

Does lactic acid cause muscle fatigue?

Conventional wisdom says that lactic acid is the root of all athletic evil. When you're exercising hard enough to go into "oxygen debt," it's lactic acid that makes your muscles burn and eventually forces you to stop, and it's leftover lactic acid that makes you stiff and sore the next day—or so exercise physiologists believed for nearly a century. We know now that this is wrong. Lactic acid is in fact a crucial fuel for your muscles, not a painful waste product. (Actually, an ion called "lactate" is what's typically found in the body, which can combine with a proton to form lactic acid.)

The roots of the lactic acid myth go back to experiments with isolated frog muscles in 1907. When researchers applied a shock to the muscles (which were disconnected from the frog's bloodstream and thus had no source of oxygen), they found that lactate was produced. However, when they repeated the experiment with oxygen supplied, the lactate disappeared. Over the next few decades, physiologists developed the hypothesis that muscles produce lactate when they're forced to contract without oxygen and that the accumulating acidity is what causes muscular fatigue.

These ideas weren't challenged until a series of experiments in the 1970s by George Brooks of the University of California, Berkeley, and it wasn't until the past decade that his views gained widespread acceptance. Brooks showed that you don't produce lactate only when you're in oxygen debt. In fact, you're constantly converting your carbohydrate stores into lactate, even when you're at rest. About half of this lactate is then immediately converted into ATP, the basic fuel for muscular contractions. The proportion of lactate used in this way hits 75 to 80 percent when you're exercising, since it doesn't require oxygen. The rest goes into the bloodstream and is used to fuel the heart or is converted by the liver into glucose (another source of energy for working muscles).

Untangling the intricate details of which cells produce and consume lactate and how your body maintains that balance remains an area of active research. But the practical implications are clear: studies have found that endurance-trained athletes produce roughly the same amount of lactate as untrained subjects, but they use it as fuel far more efficiently—that's why lactate levels in their blood don't rise as quickly. This means that, even though we've had the science backward all this time, we've somehow gotten the practical advice for training right. The goal of exercising at or just below your "lactate threshold" (see p. 66) makes perfect sense—but the aim is to teach your body to consume lactate more quickly, not to avoid "poisoning" your muscles with too much lactate.

If that's the case, what actually causes fatigue in your muscles? One recent theory advanced by researchers at Columbia University is that exhausted muscles begin to leak calcium, which reduces the force of muscle contractions. While this may be part of the puzzle, it's unlikely that there's one single factor that causes fatigue. And even though lactate is a valuable fuel, it's still possible that increasing acidity in your muscle tissue can interfere with muscle contraction and cause the acute discomfort you feel. As for post-exercise muscle soreness that crops up a day or two later, there was never any reason to blame that on lactic acid—which, as it turns out, returns to normal levels in your blood less than an hour after hard exercise.

Why do I get sore a day or two after hard exercise?

Scientists have made great progress in discovering what *doesn't* cause soreness after exercise. It's not a low-grade persistent muscle spasm (see p. 131), and it's not an accumulation of lactic acid (see p. 59). Instead, most researchers now agree with a theory first proposed over a century ago that blames post-exercise soreness on microscopic "tears" in your muscles. But that leaves an important mystery: if muscle damage is the cause, why does the pain peak 24 to 48 hours after you stop exercising?

You may know from bitter experience that the harder you work out, the more likely you are to be sore afterwards. But intensity isn't the only factor, as numerous experiments with hill running and stair climbing have shown. For example, Swedish researchers compared three groups of volunteers who ran for 45 minutes on a treadmill, either on an uphill slope of four degrees, a downhill slope of four degrees, or a downhill slope of eight degrees. Even though the uphill group had to work the hardest to maintain pace, it was only the downhill groups that developed "delayed-onset muscle soreness," or DOMS.

The reason is that downhill running involves "eccentric" muscle contractions, which occur when the muscle is trying to shorten but is being forced by an external load to lengthen. Typical examples include lowering the weight in a biceps curl or the braking action of your quadriceps (front upper leg) muscle as you run downhill. During eccentric contractions, your muscle filaments are stretched to their limits—and sometimes beyond. The resulting damage effectively weeds out the weakest links in your muscles, so that they will be stronger once they're repaired.

Ironically, it's the repair process, rather than the damage itself, that is thought to cause pain in the day or two after exercise.

The body sends cells called neutrophils and macrophages to clear out the damaged tissue and mobilizes a host of other types of cells to begin the rebuilding process. The outer membranes of nearby muscle cells get damaged in the process, allowing fluid to rush in and cause the muscles to swell. Meanwhile, another substance called bradykinin is released by the damaged muscle, which, after a delay of about 12 hours, causes an increase in levels of "nerve growth factor" that lasts for about two days. Nerve growth factor, which is associated with chronic pain conditions, makes your nerve endings more sensitive—so that any movement of your inflamed muscles presses against these hypersensitive nerves and causes pain. Recent studies in Japan and elsewhere have suggested that nerve sensitivity caused by bradykinin is enough to explain the delayed response of DOMS, and any inflammation is purely coincidental. This debate hasn't yet been resolved.

Still, the practical message is clear: once you've done the crime, you'll have to serve the time. The pain results from muscle damage, and once the workout is over you can't "undamage" the muscles, despite the promises of various lotions, creams, and pills. The good news, though, is that the damaged muscle comes back stronger once it's repaired. In fact, without this damage-repair cycle, you wouldn't get any benefit from training—so ideally, you want your workout to fall in that sweet spot where you're doing enough microscopic damage to stimulate adaptation, without doing so much damage that you have to skip the next few workouts. As you weed out the weak muscle fibers, you'll become less and less susceptible to DOMS. And you don't have to suffer the full effects of DOMS to get this protective effect—so a little moderation when you're getting back into working out or trying a new exercise can allow you to avoid DOMS entirely.

What is "VO$_2$max" and should I have mine tested?

VO$_2$max is a term that surfaces whenever feats of great endurance like the Tour de France are in the news. It refers to "maximum oxygen uptake," the greatest amount of oxygen you're able to deliver to your muscles when you're exercising as hard as you can. The more oxygen you can process, the faster you'll go—which is why many athletes seek out the VO$_2$max testing available at universities and labs for around $100 to $150.

Researchers usually measure VO$_2$max with a progressive test on a treadmill or stationary bike that starts at an easy pace, gradually accelerates, and finishes at "voluntary exhaustion" after 10 to 12 minutes. The amount of oxygen you're using, measured with tubes attached to your mouth, increases steadily the faster you go and usually starts to plateau shortly before you stop—the plateau signals that you've reached your VO$_2$max. Some researchers believe this occurs when your heart is pumping oxygen-rich blood to your muscles as fast as it can; others believe the limit lies in the muscles themselves. A more recent theory suggests that the limits aren't physical at all and in fact are dictated by the brain's instinct for self-preservation (see p. 57).

There's no doubt that elite endurance athletes tend to have higher VO$_2$max values than weekend warriors, but not for the reasons many people think. There's a common misconception that your heart will be able to beat faster—and thus deliver more oxygen—as you get fitter. In fact, top athletes tend to have lower maximum heart rates than non-athletes. Instead, they have bigger, more flexible hearts that eject more blood with each powerful stroke. The volume of blood pumped by an athlete's heart might jump from 5 liters per minute at rest to more than 30 liters per minute at top speed—twice the level that an untrained person could reach. (The highest measured cardiac output is 42.3 liters per minute, in a world-class orienteer.)

The differences in VO_2max are partly due to simple genetics and partly to hard training. A typical adult male might expect to have a VO_2max value of between 30 and 40 milliliters of oxygen per minute for each kilogram of body weight (mL/min/kg), while an adult woman would expect 25 to 35 mL/min/kg. Lance Armstrong, on the other hand, had a VO_2max of at least 85 mL/min/kg during his streak of Tour de France victories, according to University of Texas exercise physiologist Edward Coyle. "We estimate that if Lance were to become a couch potato, his VO_2max would not decline below 60 mL/min/kg," Coyle noted in a summary of the research. "Furthermore, if the normal college student were to train intensely for two or more years, his VO_2max would not increase above 60 mL/min/kg."

Despite this high value, it would be a mistake to conclude that Armstrong's victories were the result of his VO_2max, since many of his competitors had similar values. Coyle believes that Armstrong's success can be explained by an 8 percent increase in efficiency—how fast he was able to bike using a given amount of oxygen—between 1992 and 1999, though other scientists have disputed Coyle's measurements. One thing they all agree on (luckily for sports fans) is that no combination of lab measurements is capable of predicting in advance who will win a race.

So other than satisfying curiosity, what does testing your VO_2max tell you? Comparing the results of several tests over time does allow you to monitor whether you're improving or not—but then again, so does entering races. For most athletes, experts recommend lactate threshold testing (p. 66) as a more useful training tool.

VO₂MAX AND LACTATE THRESHOLD

Exercise testing is typically performed on a stationary bike or treadmill, starting at a gentle pace that increases progressively. Tubing attached to the head measures the gases entering and leaving your mouth.

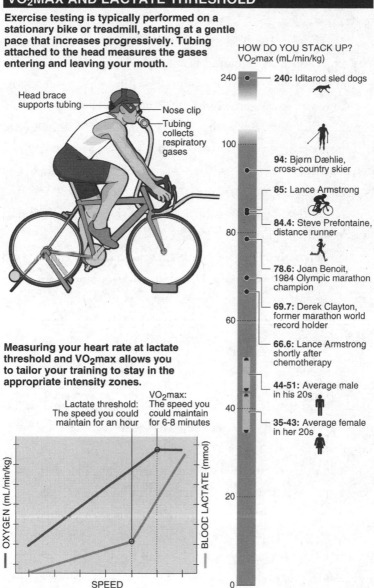

Head brace supports tubing

Nose clip

Tubing collects respiratory gases

HOW DO YOU STACK UP?
VO₂max (mL/min/kg)

240

240: Iditarod sled dogs

100

94: Bjørn Dæhlie, cross-country skier

85: Lance Armstrong

84.4: Steve Prefontaine, distance runner

80

78.6: Joan Benoit, 1984 Olympic marathon champion

69.7: Derek Clayton, former marathon world record holder

60

66.6: Lance Armstrong shortly after chemotherapy

44-51: Average male in his 20s

40

35-43: Average female in her 20s

20

0

Measuring your heart rate at lactate threshold and VO₂max allows you to tailor your training to stay in the appropriate intensity zones.

Lactate threshold: The speed you could maintain for an hour

VO₂max: The speed you could maintain for 6-8 minutes

OXYGEN (mL/min/kg)

BLOOD LACTATE (mmol)

SPEED

What is "lactate threshold" and should I have mine tested?

Scientists are still arguing about the physiology of the lactate threshold and how it should be defined, but the basic concept is straightforward. If you're running or biking along at a slow pace and you're reasonably fit, you should feel like you could continue for hours. If you're going very fast, you'll feel like you need to stop or slow down within a few minutes. Somewhere between those two paces, there's a point beyond which you begin burning energy at a rate that you can't sustain, and that point is marked by a dramatic increase in the rate at which lactate accumulates in your blood.

This threshold typically corresponds to the pace you could maintain for about an hour, and it's accompanied by other physiological indicators; for instance, you'll start breathing much more heavily, which is why the "Talk Test" (see p. 80) can be used as a rough guide to find your threshold. The pace you're going when you hit that threshold is the best indicator scientists have to predict how you'll fare in a race, and it's also a valuable guide for figuring out how fast you should go in workouts—which is why many athletes use regular lactate testing to monitor progress and guide training.

Scientists originally believed that lactate was a harmful waste product that caused pain and fatigue—but they were confusing correlation with causation (see p. 59). Lactate levels rise when your muscles are in "oxygen debt," or forced to burn energy less efficiently because they can't get enough oxygen— but lactate is actually a fuel rather than a waste product. Still, you can use the rise of lactate levels in your blood as a rough indicator of when you're making the transition from relying mostly on aerobic metabolism (your muscles are getting enough oxygen to keep going) to anaerobic metabolism (they're no longer getting enough, so you can't continue indefinitely).

Like the VO_2max test, a lactate threshold test involves exercising on a treadmill or stationary bike with the speed gradually increasing. Lactate tests typically last between 20 and 60 minutes, with the speed staying constant for about five minutes at a time. At the end of each stage, a pinprick of blood is extracted from the fingertip or earlobe. Absolute values of lactate aren't very significant, because the number can vary depending on how the measurements are made or even what you've eaten beforehand. Instead, the significant result is your speed (and heart rate) when lactate levels start to increase dramatically. This is your lactate threshold.

A 2009 review in the journal *Sports Medicine* surveyed 32 studies of the link between lactate threshold and racing performance in running, cycling, race walking, and rowing. The results showed that lactate threshold was far more accurate than VO_2max at predicting results, accounting for between 55 and 85 percent of variance in running distances from 800 meters to the marathon.

Lactate threshold is also more sensitive to training than VO_2max, so it's an ideal tool to monitor training. Adam Johnston, a coach and the director of the Endurance Lab in Toronto, has the athletes in his WattsUp Cycling program take lactate tests every four months. "It's very empowering for the athlete to see the definite differences after four months of training," Johnston says. "And it also helps to detect when things aren't working, so we can make changes."

Of course, there are many other ways to monitor your training, starting with the humble stopwatch. But if you crave objective data, love cutting-edge technology, and have always wondered how you stack up with Lance Armstrong, the tests are easily available. "There's a misconception that it's only serious, elite athletes who do this," Johnston says. "But we see a huge range of people. It's for anyone who wants to be fit, but also has a performance goal they're working toward."

How can I avoid muscle cramps?

Researchers from Brigham Young University reported an odd finding in 2010: drinking a quarter-cup of pickle juice gets rid of muscle cramps within an average of about 85 seconds, 45 percent faster than they'd disappear on their own. What's interesting about this is not the promise of a "cure"—85 seconds is still a long time to be in pain—but the question of why it works. For years, we've been told that muscle cramps result when you become dehydrated and lose too many electrolytes, the salts that are carried away in your sweat. But this tiny amount of pickle juice isn't enough to restock fluid or electrolyte levels in the body, and the effect occurs long before the pickle juice could possibly be absorbed from the small intestine.

Instead, researchers are now considering the possibility that cramps are a phenomenon related to "altered neuromuscular control," stemming from several factors, including fatigue, muscle damage, and genetics. The new theory doesn't offer any quick fixes, but it suggests that proper training and pacing could help minimize your risk.

The first studies of muscle cramps date back over a century, to studies of miners and steamship workers laboring in hot, humid conditions. The idea that replacing the water and salt lost in sweat would prevent cramps developed from these early observations—but no controlled trial has ever managed to show that it actually works. In fact, several studies comparing cramp-prone

Ironman triathletes to their non–cramp-prone peers by research-ers at the University of Cape Town have found that hydration and electrolyte levels in the two groups are almost indistinguish-able before and after the race. And a forthcoming study from the Brigham Young group forced volunteers to exercise until they lost 3 percent of their body mass through sweat but found no change in their susceptibility to electrically stimulated cramps.

The neuromuscular cramp theory was first proposed in 1997 by University of Cape Town sports physician and researcher Martin Schwellnus, to explain simple observations like the fact that the muscles affected are usually those that have been working hardest. "If it's a systemic problem like dehydration, then why doesn't the whole body cramp?" he asks. In addition, he adds, sports doctors working in the medical tent at athletic events have long known that the best way to relieve a cramp is to stretch the affected muscle—another hint that the problem is local rather than general.

Your muscles are always held in a delicate balance between two types of reflex signal carried by the nervous system: an ex-citatory input that encourages them to contract, and an inhib-itory input that encourages them to relax. Schwellnus believes that several factors associated with exercise can upset this bal-ance, increasing the signal from the excitatory reflex and lower-ing the signal from the inhibitory reflex. When this occurs, he says, "the muscle gets twitchy." If the imbalance persists, the muscle will eventually contract in a full-blown cramp.

Factors that can affect the balance and make your muscles twitchier include fatigued or damaged muscle fibers, which would explain why cramps generally occur in the hardest-working muscles. Experiments with rats suggest the vinegar in pickle juice can alter these reflex signals, and the fact that cramping often runs in families suggests a genetic component. In contrast, stretching a muscle triggers the inhibitory reflex, which explains why it's a painful but effective way of ending a cramp.

Interestingly, Schwellnus's study of triathletes found that those who developed cramps had set higher pre-race goals and started at faster paces relative to their previous best times compared to non-crampers. And in a further study that hasn't yet been published, he found that crampers tend to have trained more in the final week before the race and to have elevated blood levels of enzymes related to muscle damage before they start.

These lessons—train sufficiently, set realistic goals, and rest before races—won't stop every cramp, but they may reduce your risk. Still, the science remains hotly disputed, so if eating an electrolyte-rich pre-workout banana has kept you cramp-free so far, don't stop now. And if you haven't tried pickle juice, you might as well give it a shot!

What's happening when I get a stitch?

The stitch is one of the most familiar afflictions of exercise. It's mentioned in the works of Pliny the Elder (who recommended "the urine of a she-goat, injected into the ears" in certain cases) and Shakespeare. But it has been virtually ignored by modern sports scientists, as Australian researcher Darren Morton noted in a recent *British Journal of Sports Medicine* review. Morton describes the pain as "sharp or stabbing when severe, and cramping, aching, or pulling when less intense." Until a series of studies was conducted over the past decade, no one even knew what the stitch (or side-cramp, or "exercise-induced transient abdominal pain") was, let alone what was causing it. Was it a cramp or spasm of the abdominal or respiratory muscles? Constricted blood supply to the diaphragm? Too much jostling of the ligaments supporting the internal organs?

Morton and his colleagues have carefully ruled out these popular theories one by one. By feeding volunteers carbonated, high-sugar soft drinks and then putting them on a treadmill, they've learned how to provoke stitches with reasonable reliability. They used electrodes to monitor abdominal muscle activity during stitches and found no change, ruling out abdominal muscle spasm. They analyzed breathing parameters during stitches and again found no change, ruling out respiratory muscle spasm. The wide variety of stitch locations, including low in the abdomen, rules out the diaphragm. And the occurrence of stitches during non-jolting activities like swimming argues against the ligament theory.

So what's left? There's still no definitive answer, but new studies have offered some revealing hints. Doctors have sometimes seen patients with stitch-like pain caused by brain lesions or by compressed nerves in a certain part of the thoracic (upper-back) spine. Morton and his team found that they could induce a stitch by pressing on this area of the spine, suggesting that there's a neural component. And in a 2010 study, they found a link between the degree of spine curvature in 159 volunteers and their susceptibility to stitches. The greater the kyphosis (curvature of the upper back) and lordosis (curvature of the lower back), the more severe the pain of the stitch.

The actual pain, the researchers argue, is best explained as irritation of something called the "parietal peritoneum," the outer layer of the membrane that lines the abdominal cavity. Friction between layers of the peritoneum could be caused by a full stomach pressing from the inside or the constrained shape of a rib cage with excess spine curvature. Notably, the nerves of the peritoneum are connected through the same region of the thoracic spine that can trigger stitches—and other peritoneal nerves are linked to the shoulder, where some people experience stitch pain.

This suggests that, if you're prone to stitches, you should avoid filling up your stomach with food or water right before exercising. It also suggests that postural exercises to limit spine curvature could be helpful. Since other studies have found that consuming high-sugar beverages before exercise makes stitches more likely, it makes sense to avoid them and to experiment with what other foods may be triggers for you. Finally, the clearest correlation they've found is with age: the younger you are, the more likely you are to get a stitch. So if all else fails, time may cure you.

At what time of day am I strongest and fastest?

One of the hottest controversies leading up to the Beijing Olympics in 2008 was the decision to switch swimming and gymnastics finals from their traditional evening slots to early morning, so that they could be broadcast live in the United States. To understand why many athletes were outraged, imagine trying to run a mile all-out or bench-press your maximum weight—at 3 a.m. It's a safe assumption that your performance wouldn't measure up to your usual midafternoon abilities. These fluctuations, related to patterns of sleep and waking, are easy to understand, but they're only part of the story. Your body contains an internal clock governed by a region of the brain called the hypothalamus, which regulates temperature, hormone secretion, sleep, and feeding cycles with a period of roughly 24 hours. As a result, your body experiences subtle

changes throughout the day—even while you're wide awake—
that influence your physical capabilities.

Researchers in France and Tunisia performed careful studies of the hourly variation in a 30-second all-out cycling bout
called the Wingate test. They found that peak and total power
hit a maximum at about 6 p.m., with values between 8 and 11
percent higher than the low point at 6 a.m. Other studies have
made similar findings about tests of back muscles, arm muscles, vertical and broad jumping, and anaerobic power, with the
peaks always within a few hours of 6 p.m. Tests of sports as
varied as running, swimming, football, badminton, and tennis
have yielded similar findings.

Spanish researchers were able to shift the peak performance
time of elite sprinters forward or back by two hours simply by
adjusting sleep-wake and meal times by two hours. But it's not
clear whether performance was related to total time awake,
or whether the change in meal and sleep times succeeded in
changing the timing of the body clock itself, leading to other
changes in the body's chemistry.

Many researchers believe that body temperature, which increases by about 1.8°F (1°C) over the course of the day, could
be the key factor. Increased core temperature could lead to
looser muscles, faster metabolic reactions in the body, and faster
transmission of nerve signals. A 2010 study by researchers in
Guadeloupe found no difference in a series of jumps, squats,
and cycle sprints performed either between 7 and 9 a.m. or between 5 and 7 p.m. The reason for the difference compared to
studies in colder climates, they suggested, could be that the consistently warm and humid climate in Guadeloupe provided a
"passive warm-up" that made the slight circadian increase in
body temperature irrelevant.

These findings may be slightly worrying if you're preparing for a competition that takes place at an unusual time, like a 10K or marathon that starts at 7 a.m. Fortunately, several studies have suggested that training has time-specific benefits. A 1989 University of Georgia study, for example, found that cyclists who trained early in the morning experienced greater improvements in their oxygen use at threshold pace when they were tested in the morning; another group that trained in the evening had larger improvements in evening testing sessions. A Finnish study in 2007 found similar results for strength training and also found that early-morning workouts changed how levels of the stress hormone cortisol fluctuated throughout the day, enhancing early-morning performance. So if you want to be at your best at 7 a.m., make sure to do some training at that time.

Despite all this evidence, the best time to work out is generally still whatever time fits best into your daily schedule, since factors like sleep, stress, and fatigue will outweigh the slight boost offered by circadian rhythms. You should also make allowances for individual differences: most of these studies were performed on volunteers with "typical" sleep schedules—neither night owls nor early risers. If you happen to have, say, a particular mutation in the hPer2 gene that makes it hard for you to stay awake past 8:00 p.m., your peak performances will likely come quite a bit earlier than 6 p.m.

CHEAT SHEET: THE PHYSIOLOGY OF EXERCISE

- Your physical limits aren't defined by the failure of your muscles, heart, or lungs; instead, there's increasing evidence that "fatigue" is regulated by subconscious processes in the brain.
- Lactic acid isn't a metabolic waste product that makes your muscles burn. It's actually a useful fuel that provides energy to your muscles; the fitter you are, the more lactic acid you use.
- Hard exercise causes microscopic damage to your muscle fibers. The repair process causes swelling and hypersensitive nerve endings, leading to "delayed onset muscle soreness" (DOMS) that peaks a day or two later.
- VO_2max is a measure of the maximum amount of oxygen you can deliver to your muscles during exercise. It's a measure of aerobic fitness but varies widely even among athletes of equal ability.
- "Lactate threshold" refers to the point at which lactate begins to accumulate rapidly in your blood, indicating that you're no longer in the "aerobic" zone. It's often used to monitor the progress of training.
- Moderate exercise boosts your immune system, but very intense exercise—serious marathon training, for example—can temporarily suppress it.
- Muscle cramps are traditionally blamed on electrolyte loss through sweat, but a new theory suggests that they're caused by disrupted neural reflexes linked to muscular fatigue.
- "Stitches" are still poorly understood but may result from friction between layers of a membrane lining the abdominal cavity. Good posture reduces your chances of getting one.
- Peak physical performance for most people occurs in the late afternoon or early evening, around 6 p.m., when body temperature is highest. You can boost your performance at a particular time of day by training regularly at that time.

4: Aerobic Exercise

EXERCISE FADS COME AND GO — and sometimes come again, like the trampoline craze that flourished in the 1950s, faded from view, then roared back with the introduction of mini-trampolines in the late 1970s. But the basic elements remain unchanged: sustained rhythmic exercise is the most important element in any general fitness program, whether you call it cardio (because it strengthens your heart and circulatory system) or aerobic exercise (because it requires oxygen).

The term *aerobics* was only coined in 1968, when Air Force researcher Kenneth Cooper published a best-selling book of that name, and originally referred to any exercise with aerobic benefits (not just 80s-style classes filled with colorful spandex). That versatility is important to remember: once you learn to judge your effort appropriately, you get a good aerobic workout just about anywhere—inside or outside, at the gym or on the way to work . . . and even on a trampoline.

Why should I do cardio if I just want to build my muscles?

Slogging away on an exercise bike will never give you bigger biceps or a faster fastball, no matter how hard you pedal. So there's a temptation to focus your workout efforts on the goals you're most interested in and neglect the rest—especially if you don't particularly enjoy cardio workouts. Whether you're focused on health or athletic performance, this is a bad idea.

Aerobic fitness is often assessed by measuring how much oxygen you can breathe in and deliver to your muscles when working as hard as you can, a quantity known as VO_2max (see p. 63). You can improve aerobic fitness by doing activities that continuously use large muscle groups—walking, running, biking, step aerobics, swimming, dancing, and so on. These activities are sometimes lumped together as "cardio" exercises because they improve the health of your cardiorespiratory system—your heart and lungs.

The adaptations that result from aerobic exercise—strengthening the heart, increasing the number of small arteries that circulate blood, stimulating the growth of power-producing mitochondria in your muscles—are quite different from the adaptations produced by strength training. Building strength is primarily the result of two things: increasing the size of your muscle fibers and improving neural pathways so that commands from your brain are executed more efficiently. Both types of exercise are important, but neither is enough on its own.

Thanks to public health messages, we've all heard about the benefits of exercise: reduced risk of heart disease, diabetes, and arthritis; longer life; lower weight; less stress, less depression, and increased cognitive function. It's important to realize that it's aerobic fitness, not strength training, that is most closely tied to all these good outcomes. If the health benefits don't convince you, the performance benefits should—even in sports where you wouldn't expect it. A 2009 study of snowboarders on the World Cup circuit, for instance, found that aerobic fitness as measured on a stationary bicycle was one of the most reliable predictors of finishing position. Even golf, the butt of jokes about athleticism, shows the same pattern: a 2009 study of the physical and physiological characteristics of 24 golfers on the Canadian national team found a strong correlation between aerobic fitness in a running test and tournament performance. Whatever the activity, you can practice harder and keep performing at your best for longer if you're aerobically fit.

This doesn't mean that you're doomed to a lifetime of treadmills and stair-machines. During a competitive soccer game, for example, players work at about 70 percent of their VO_2max, well above the minimum of about 50 percent needed for an effective aerobic workout. Baseball and football, on the other hand, have too much standing around to offer much benefit. For any sport, the benefits depend on how vigorously and how continuously you play.

Some people focused on weight lifting have turned to "circuit training" as an aerobic alternative. This popular form of weight training involves moving rapidly from one strength exercise to the next with only a few seconds of rest, in an attempt to keep the heart rate high enough to get both aerobic and muscular benefits. Unfortunately, studies over the last 30 years have consistently found that circuit training typically makes people work at a little less than 50 percent of their VO_2max. Of course, this isn't an iron-clad rule. It's possible to get an aerobic workout from a circuit routine if you keep the intensity high enough and the recovery times short enough. This is the approach taken by popular all-around exercise regimens like CrossFit, and it's perfectly reasonable as long as you maintain a consistently high intensity that's nearly nonstop.

The bottom line: yes, you need to do aerobic exercise, no matter what your sports performance or health goals are. You can fit it in by doing a 10- to 15-minute stint on a cardio machine after lifting weights, by playing sports, or by setting aside a couple of workouts a week to focus on cardio. You don't have to go jogging—but it should feel like you did.

How hard should my cardio workout feel?

One of the easiest mistakes new exercisers can make is to be overzealous—to jack the treadmill up to maximum velocity until they get spat out the back, and repeat that every day until they're too discouraged and exhausted (and possibly injured) to continue. Of course, it's just as possible to be underzealous, flipping through a magazine while absent-mindedly turning the pedals of the exercise bike at a glacial pace. Ideally, you want to be somewhere between these extremes—but where, exactly?

For aerobic exercise, you can divide efforts into three basic zones on the basis of how your body reacts. The easiest is the aerobic zone, where your heart and lungs are able to deliver enough oxygen to your muscles to keep them functioning. In contrast, the hardest, or anaerobic, zone is where your muscles can't get enough oxygen. In between is the threshold zone, which is marked by a rapid accumulation of lactate in your blood as your muscles begin to cope with insufficient oxygen (see p. 66). (The terminology, and indeed the science underlying these descriptions, is still a topic of active research and debate. But all sides agree about the practical training advice, so we'll stick with "aerobic," "threshold," and "anaerobic" for now.)

How much time you spend in each of these zones depends on your goals and preferences, but a good general guideline for how to allocate your workout time over a typical week is to aim for 70 percent aerobic, 20 percent threshold, and 10 percent anaerobic. These ratios are based on studies of elite endurance athletes, who attempt to balance the hard training that brings the largest and swiftest gains in fitness with easier efforts that allow them to recover while continuing to improve. For many novices, the high proportion of "easy" efforts is a surprise—but it's simply not possible to hammer all the time.

There are many sophisticated ways of figuring out what zone you're in. Exercise labs, for instance, measure the lactate in your blood at different intensities to precisely determine the speed or heart rate at which your threshold occurs. Heart rate monitors can also be used to monitor training intensity, though it's important to understand their limitations (see p. 81). Surprisingly, going by feel can be just as accurate as the high-tech approaches, and it has the added benefit of forcing you to pay attention to your body's signals so that you'll know when something is amiss.

A 1987 study at the University of Liverpool found that instructing cyclists to pedal with vague descriptions like "somewhat hard" or "hard" was just as good as heart rate at producing a repeatable effort. Since then, numerous studies have confirmed that "perceived exertion" is a reliable way of gauging intensity. For example, on a scale of 1 to 10, the aerobic zone corresponds to about 3 (moderate); threshold is 5 to 6 (hard); and anaerobic is 7 to 9 (very hard). Only about 10 percent of people struggle with perceived effort, according to University of Wisconsin-La Crosse researcher Carl Foster—mostly control-oriented people (often lawyers or surgeons) who don't like to admit anything is difficult. "They're on a treadmill saying, 'This is easy, this is pretty easy, this is sort of moderate'—and then they're going backward off the treadmill," he says.

An even simpler approach is called the Talk Test. If you're able to speak in full sentences—out loud, not under your breath—you're in the aerobic zone. Once you hit the threshold, you'll start to breathe much harder and only be able to speak in short phrases. And when you can only gasp out a word or two at a time, you're in the anaerobic zone. What Foster has found over the years is that, left to their own devices, athletes tend to go too hard on their easy days. (The reverse can also be a problem—note that the aerobic zone should feel "moderate," not "easy.") By paying close attention to your effort—and by talking out loud, even if you're alone—you'll be able to avoid those pitfalls.

Zone	AEROBIC	THRESHOLD	ANAEROBIC
Training time	70%	20%	10%
Heart rate	Below 80% of max	80–90% of max	Above 90% of max
Talk Test	Complete sentences	A few words at a time	Single words
Workouts	20–60 minutes steady	Surges of 3–10 minutes	Short bursts of 0.5–3 minutes

How do I determine my maximum heart rate?

You've seen the charts on posters at the gym and on the control panels of cardio machines: keep your heart rate in the right zone—warm-up, "fat-burning," aerobic, and so on—to get the most out of exercise. There's just one problem: in order to exercise at, say, 75 percent of your maximum heart rate, you need to know what that maximum is. And despite what it may claim, your treadmill really doesn't know you that well.

The conventional wisdom, used by even the fanciest cardio machines, is that maximum heart rate is 220 minus your age. It's simple, easy to calculate, and—more often than not—wrong. "It turns out there's not much scientific evidence for that formula," says Hirofumi Tanaka, head of the University of Texas's

Cardiovascular Aging Research Laboratory. In fact, it started as a simple rule of thumb based on a few studies in the early 1970s that included smokers and patients on heart disease medication, and almost no subjects over the age of 55. A better formula, Tanaka has found, is 208 minus 0.7 times your age. But even that provides only a rough average, because there's so much natural variation in heart function. For about one in every three people, the formula will be off by more than 10 beats—enough to put you in a completely different exercise zone. In addition, some recent studies have suggested that aerobic training can lower your maximum by as much as 10 beats per minute as your heart gets stronger.

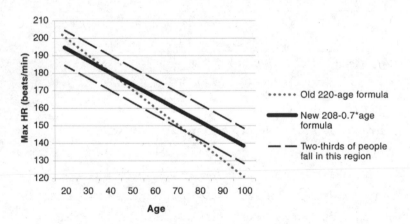

The only way to reliably determine your personal max, it turns out, is to actually get your heart pumping that fast. That typically requires a cardiac stress test, usually performed in a lab with a treadmill. You can also get a decent estimate by wearing a heart rate monitor for a 15- to 20-minute run in which you gradually accelerate from a jog to an all-out sprint for the last few minutes. Ideally, finish with a couple of laps on a local track, advises veteran Atlanta-based running coach Roy Benson: "After every hundred meters, look at your heart rate and accelerate as hard as you can," he says. "You're looking for your heart rate to peg at an upper limit." You could do this at a local 5K race, where the crowds and competition will help motivate you to reach your true maximum—but the key is beginning gradually, so that your legs don't fail before your heart maxes out.

Knowing your true max allows you to get a more realistic sense of what training zone you're in. Still, there are other pitfalls you need to watch out for. In hot and humid conditions, sweating can reduce your blood volume and cause "cardiac drift": your heart will beat faster even though your effort stays low. In cool and dry conditions, the opposite can occur, keeping your heart rate artificially low even when you're going at the desired pace. That means that, while the heart rate monitor is a useful tool, the final judge of training effort should always be how you feel.

What's the best way to breathe during exercise?

Birds do it, and so do horses. So it seems natural to expect that humans would synchronize their breathing with rhythmic activities like running, cycling, or rowing. Sure enough, decades of studies have found links between stride rate and breathing rate for both novices and experts, at slow and fast speeds, in many different activities. Some studies suggest this unconscious synchronization makes your movement more efficient—but the latest research suggests that trying to force yourself to breathe in a certain pattern can backfire.

Horses maintain a fixed one-to-one ratio between strides and breaths because their lungs and breathing muscles are shaken rhythmically by the impact of their hooves on the ground.

Flapping wings put birds under similar constraints. Humans, on the other hand, walk upright, so the jarring impact of each foot-strike doesn't directly interfere with breathing muscles. Still, a series of studies in the 1970s showed that if you put subjects on a treadmill or exercise bike, some (but not all) naturally fall into a pattern where their breathing rate is synchronized with their cadence. The ratio of full strides (i.e., counting each time the right foot hits the ground) to full breathing cycles (i.e., counting each exhale) varies widely among subjects, with common observations of 1:1, 2:1, 3:1, 3:2, 4:1, and even 5:2. The most common among runners is two stride cycles for each breath.

A Swiss study in 1993 found that runners seem to burn slightly less energy when their breathing is coordinated with their stride rate—a finding that spurred some coaches to encourage their runners to focus more on their breathing. Numerous studies since then have explored whether synchronized breathing makes movement feel easier or burns less energy, with conflicting results. After several decades of research, the failure to demonstrate any clear link between synchronized breathing and efficiency suggests that if there is any effect, it's too small to be of practical significance. Still, the idea that there's a "correct" breathing pattern has persisted ever since.

More recent research has suggested that trying to consciously control your breathing is quite different from letting your breathing adopt a rhythm subconsciously, and it may even have negative effects. A 2009 study from the Institute of Sports Science in Münster, Germany, had runners focus their attention on either their surroundings or their breathing. When they focused on their breathing, the subjects took deeper breaths and slowed their breathing rate from 37 breaths a minute to just 30. The result: they burned almost 10 percent more energy compared to when they simply let their mind wander. The researchers conclude that we're pretty good at picking the right breathing rate without thinking about it, so we just make things worse when we try to interfere.

Still, there are some useful breathing tips that new exercisers should know. For instance, it's easier to get lots of oxygen in by breathing through your mouth and nose rather than either one on its own. And synchronizing your breathing is important if you're lifting weights: exhale as you lift and inhale as you release, making sure not to hold your breath. But in general, if you find yourself panting uncontrollably as you start a new cardio activity, it's far more likely because you're starting too fast than because you're breathing wrong. Slow down, enjoy the scenery around you, and let the breathing take care of itself.

Will running on hard surfaces increase my risk of injuries?

The conventional wisdom is deeply entrenched: constant pounding on hard surfaces will beat up your legs and lead to injury. But proving this has turned out to be remarkably tricky—in fact, a growing number of studies suggest that, despite our intuition, we're able to automatically adjust our running stride so that hard and soft surfaces administer roughly the same shock to the body. Instead, researchers suspect that the crucial difference between running surfaces may be how smooth or uneven they are.

The surprising idea that your body can make adjustments for different running surfaces dates back to studies in the 1990s. Scientists found that when they varied the stiffness of a running surface, runners adjusted the effective stiffness of their legs in the opposite direction—by bending their knees slightly more or less and by tensing their muscles—so that their total up-and-down motion remained perfectly constant.

In support of this notion, a 2002 study by Mark Tillman of the University of Florida, using force-sensing shoe inserts, found no difference in the in-shoe forces felt by runners on asphalt, concrete, grass, and a synthetic track. This result has been called into question by a 2010 Brazilian study that found 12 percent more pressure with each foot-strike when running on asphalt compared with grass. Still, even that difference is surprisingly small given the large difference in the softness of the two surfaces.

Forces and pressures are only part of the picture, though, Tillman cautions. For example, slight differences in knee angle as your leg adjusts to a different surface could theoretically translate into greater likelihood of injury on one surface compared with another. But the simple picture—harder surface leads to more pounding leads to injury—isn't supported by the existing evidence.

It may be that our focus should be on how smooth the surface is rather than how hard it is, according to Stanford University biomechanics researcher Katherine Boyer. Flat, paved surfaces will result in every stride being almost identical, so your muscles, joints, and bones are stressed repeatedly in exactly the same way throughout the run. That sets the stage for overuse injuries like shin splints, which typically occur when you try to increase your training too quickly. (The same factors would presumably be present on treadmills, though researchers have yet to investigate this.) On unpaved surfaces, in contrast, no two steps are the same, which provides slight variations in the impacts on your body. Too much unevenness, though, carries risks such as a turned ankle. "The key is to find the balance between stress and overstressing the system," Boyer says.

ADAPTING TO DIFFERENT RUNNING SURFACES

When you're running, your legs function as springs to absorb the impact of your foot-strike. The ground provides another spring to absorb some shock. In a surprising 1998 paper, researchers at the University of California, Berkeley, found that when the springiness of the ground was changed, runners automatically adjusted the springiness of their legs by altering knee angle and muscle tension. The result: their overall up-and-down motion stayed the same regardless of the surface.

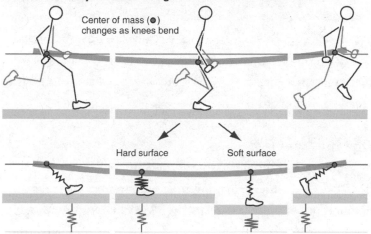

Center of mass (●) changes as knees bend

Hard surface   Soft surface

CHOOSE YOUR SURFACE

1. DIRT AND GRAVEL: A well-maintained gravel road or trail is the best surface you can find: firm and flat, but yielding. Dirt is also good, but watch for ruts and holes.

2. CONCRETE: The hardest surface of all, many runners try to avoid it. The real problem may be that it's too smooth, so every step is the same. Use in moderation.

3. ASPHALT: It's everywhere, so it's convenient. But cambered roads, which slope away from the center for drainage, put an uneven strain on your joints. Alternate sides if traffic permits.

4. GRASS: Running on the courts at Wimbledon would be heavenly. Everywhere else, clumpy and uneven grass raises the risk of twisted ankles, so stick to well-trodden paths.

5. SYNTHETIC TRACK: Rubberized tracks are softer than asphalt and well-measured (usually 400 meters), but always turning in one direction raises injury risk. Alternate directions periodically.

So far, on-the-ground studies such as one from 2003 that followed 844 runners preparing for the Vancouver Sun Run 10K have failed to find any association between running surface and injury rate. That doesn't mean no relationship exists, just that it's more complicated than we initially suspected. One thing we do know is the principle of specificity: if you do all your training on one surface, your body may not be adequately prepared to run on other surfaces. So don't prepare for a marathon by running only on grass, and don't prepare for a backcountry trail race by running only on roads.

Do I run "wrong"?

In some ways, running has more in common with basic functions like eating and breathing than with more technical sports like golf or swimming. As kids, we learn to run with no special instruction, just as our ancestors have for millennia. Still, it's clear that some people run more smoothly than others. As a result, an industry has emerged promising to teach people "correct" running techniques like the Pose Method or Chi Running.

Two questions arise. First, is it possible to change the way you've run all your life after a few weeks of workshops? And second, does it do any good?

Though there's a shortage of long-term research in this area, the answer to the first question appears to be yes. A 2004 study by researchers in South Africa put 20 runners through an intensive one-week program to learn the Pose technique, which aims to place the body in an S-shaped pose as each foot hits the ground. Sure enough, running with the new technique produced shorter stride lengths, less vertical oscillation, and less power absorbed by the knee. Whether these changes are maintained—and eventually start to feel "natural"—has yet to be tested.

The second question is more controversial. You can imagine that the reduced load on the knee might translate into fewer injuries, though that wasn't demonstrated. The problem is that these forces don't simply disappear—they're transferred elsewhere, in this case to the ankle. In an interesting postscript, one of the scientists involved in that study, Ross Tucker, revealed in 2008 on his blog, *The Science of Sport*, that 14 of the 20 runners in the study suffered from calf or Achilles tendon problems in the two weeks after the study ended.

In 2005, another study of the Pose technique, this time at Colorado State University, confirmed once again that runners could learn to use a shorter stride length with less vertical oscillation. This time, eight volunteers were given 12 weeks of instruction, and their running economy—a measure of how much oxygen is consumed at a given speed—was measured before and after. The result was a statistically significant worsening of running economy by about 8 percent. There was one interesting anomaly in the Colorado State study: the only subject who improved his running economy happened also to be the least experienced runner in the group, and he started the study with the longest stride. One possible conclusion is that the other, more experienced runners in the study had already developed optimal stride patterns based on trial and error, which is why they got worse when they tried to learn a new technique. The more general conclusion is that the Pose method is good for some people and bad for others.

Some researchers are looking at simpler interventions, like taking shorter, quicker steps. Bryan Heiderscheit, the head of the University of Wisconsin's running injuries clinic, noticed that the less-experienced runners among his patients were overwhelmingly prone to "overstriding"—reaching too far forward with each step so that the heel comes crashing down well in front of the body. "As they make the transition from fast walking to slow running, they tend to keep the long stride and low step rate," he observes. In a 2010 study, he and his colleagues found that increasing stride rate by 5 to 10 percent (initially using a metronome until the new cadence felt natural) decreased the impact forces felt in the knee and hip. They're now undertaking a multi-year clinical trial to find out whether this simple change will result in fewer injuries.

For now, there simply isn't enough experimental data to draw firm conclusions about what happens when you try to learn to run "better." It's clear that, with sufficient instruction and practice, you can alter the way you run. But it may be that the best way to learn to run is simply to run, relying on your body's innate search for efficiency to find the right technique for your body.

What's the best way to run up and down hills?

Running uphill is tough on the lungs, and running back down is tough on the legs. Either way, hills can derail an otherwise pleasant run if you're not prepared for them. Fortunately, a team of Australian researchers armed with the latest technology has come up with some valuable guidance. In a 2010 study, they sent a group of runners out on a hilly six-mile course while wired with a portable gas analyzer to measure oxygen consumption, a GPS receiver to measure speed and acceleration, a heart rate monitor, and an "activity monitor" to measure stride rate and stride length. The results suggest that most runners make two key mistakes: they try to run too fast on the uphills, and they don't run fast enough on the downhills.

When you're running on flat terrain, your speed is generally limited by the ability of your heart and lungs to transport oxygen to the muscles in your legs. If you try to maintain the same speed while hauling your body up a hill, you'll quickly notice that you're breathing harder because you're consuming more oxygen. The problem with this approach is that, once you get to the top of the hill, you'll need time to recover from this extra effort. In the study, runners took an extra 78 seconds on average to regain their initial speed after cresting a hill—a delay that wipes out the benefit of pushing hard up the hill, lead researcher Andrew Townshend of the Queensland University of Technology says. "Based on our results, we suggested that a small decrease in speed on the uphill may be more than compensated for by a quicker return to faster running speeds on the subsequent level section," he says.

Surprisingly, the opposite was true on downhill sections. Because of the jarring impacts involved in running downhill, most of us simply can't run fast enough downhill to be limited by oxygen. The practical tip: when you get to the bottom of a hill, focus on maintaining your momentum (and higher speed) until your breathing forces you to slow down again. The downhill results were much less consistent among subjects than the uphill and level sections of the experiment. Some people were able to run far closer to their aerobic limits than others, gaining valuable time without getting significantly more tired. This suggests that downhill running is a skill you can acquire through practice.

Of course, there's a reason we tend to back off when running down hills: it's hard on the legs and raises injury risk. For that reason, it's best to limit downhill training to short sprints on a fairly gentle grade—a technique that's also used by sprinters and football players to improve their sprint speed. A 2008 study from Marquette University found that a 10 percent grade

(5.7 degrees) is the ideal gradient to maximize your speed in 40-yard sprints.

While these simple tips—slow down on the ups, speed up on the downs—should help you distribute your effort more evenly during runs, you'll need to try them out to find the right balance for yourself. "The best I can suggest is that runners should practice varying their degree of effort on hills that they frequently use in training, to determine how much they should slow down to reap an overall time benefit," Townshend says. "An experiment of n=1 for all to try!"

Does pumping my arms make me run faster?

To see first-hand how intertwined your arm and leg motions are, try running with your arms pinned to your sides. It's hard! So if moving your arms less makes you slow down, does moving them more help you speed up? That's the idea when coaches tell their athletes to pump their arms while they're sprinting or approaching the end of a long race. As it turns out, though, the link between arms and legs isn't quite that simple—in fact, it's still a topic of hot debate among researchers.

To pump your arms, you need to activate your arm and shoulder muscles. But a 2009 paper in the *Journal of Experimental Biology* challenged the idea that muscles play any role in normal arm swing during walking or running. They performed a series of experiments suggesting that arm swing is passive rather than active, occurring naturally to counterbalance the rotation of your lower body. In other words, if you had rubber arms with no muscles in them, the twisting motion of your lower body would cause the arms to swing exactly the way they do in real life (other than minor details like keeping your elbow bent while running).

Support for this notion is found in a 1998 paper from Yugoslavia, in which researchers carefully measured the effects of attaching small weights to the arms and legs of sprinters. As you'd expect, even small loads of about a pound on the legs dramatically slowed sprint speed—but loads of up to 1.5 pounds on the arms produced no effect whatsoever. If arm motion was really helping to drive leg speed, you'd expect the arm weights to have an effect. The logical conclusion is that you should focus on keeping your arms relaxed so they can swing freely, rather than trying to force them to pump faster. To achieve this, coaches often recommend keeping your shoulders low and your hands unclenched.

But not everyone agrees that the arms are strictly ballast. Researchers at the University of Michigan studied patients rehabilitating from spinal cord injuries and reported an interesting observation in 2006: rhythmic arm movement seemed to help the patients as they tried to regain movement in their legs. They saw this as evidence that arm and leg movement originate in the same part of the brain, suggesting that "arm swing may also facilitate lower limb muscle activation via neural coupling." Other researchers had similarly argued that, as a legacy from our long-distant past as quadrupeds, the motion of all four of our limbs is still coordinated. To test this idea, the Michigan researchers strapped healthy volunteers into a "recumbent stepper"—basically an exercise machine that requires you to move either your arms, your legs, or both. Sure enough, they found that when subjects moved their arms back and forth, their brains were able to recruit more muscle fibers to contract in their legs than when they didn't pump their arms.

This research is still too preliminary to draw any firm conclusions. It doesn't contradict the finding that our arms swing primarily to balance the motion of our legs. But it does suggest that, when you're fatiguing at the end of a hard run and your legs are starting to fail, keeping a nice, steady rhythm with your arms just might help you keep going.

Do spinning classes offer any benefits that I can't get from biking on my own?

In theory, cycling is cycling. The pulse of the music, the exhortations of your instructor, and the presence of a group of like-minded exercisers do nothing to spin your pedals. In practice, though, the ingredients of a typical indoor cycling class somehow combine to lift workouts to heights that most participants wouldn't achieve on their own. The alchemy of group exercise is well known to runners and aerobics classes (see p. 245), but spinning has found a recipe so powerful that researchers studying it have been forced to re-evaluate their definition of "maximal" exercise—and sound a warning for beginners who may wander into a class unprepared.

The current incarnation of group indoor cycling dates back to 1987, when South African–born cyclist Jonathan Goldberg first organized training sessions in the style he later trademarked as "Spinning." These days, more than half of North American health clubs offer group cycling classes, reaching millions of participants, according to figures from the International Health, Racquet and Sportsclub Association. In a typical class, the instructor leads participants through a ride that varies dramatically in intensity, increasing and decreasing resistance to simulate hills and headwinds. Crucially, each person controls the resistance on her own bike—only the broad contours of the workout are synchronized. "Successful instructors turn out to be really good at motivating people to push harder," says Carl Foster, an exercise scientist at the University of Wisconsin-La Crosse and past president of the American College of Sports Medicine.

A few years ago, Foster and his colleagues enlisted 20 female students for a study of the physiological responses to indoor cycling, in order to investigate earlier reports that spinners could exceed their "VO_2max"—a measure of the maximum rate at which your body can send oxygen to its working muscles. The results, published in the *Journal of Strength & Conditioning Research* in 2008, confirmed that spinners were somehow exceeding the "maximum" that earlier tests had calculated for them.

Spinning doesn't, in fact, have any magical effect on oxygen circulation. The results simply indicate that people in an ordinary cycling class managed to reach higher peak intensities than they did during the rigorous progressive exercise tests that doctors and researchers use to measure VO_2max—and much higher intensities than a typical gym user slogging away on a solitary exercise bike. These peaks are held for only short periods of time, and the average intensity throughout the session is relatively moderate: typically 65 to 75 percent of maximum intensity, Foster found. This pattern of highs and lows mimics the "interval" workouts used by endurance athletes to maximize fitness.

In general, this is a good thing. But it does carry risks for new gym users aged over 40, who may have undiagnosed heart disease, Foster cautions. A simple screening tool like the Physical Activity Readiness Questionnaire, PAR-Q (see p. 17), can help determine whether you should see a doctor before you start spinning. Some clubs offer classes aimed at beginners, which is a good way to start. After you've been spinning for a few months, the greatest danger will have passed, Foster says. Then you're free to give in to what the music, the instructor, and the group are urging you to do: go all out.

Will taking the stairs make a real difference to my health?

Sprinting up the 1,576 steps of the Empire State Building, as participants in the annual Empire State Run-Up have done each year since 1978, certainly qualifies as a vigorous workout. But you don't have to work in a skyscraper—or enter stair-climbing races—to get the benefits of a stair workout. Researchers in Ireland have been studying the benefits of dashing up the stairs periodically over the course of a workday, and they've observed surprising fitness gains. "I think the key thing here," says Colin Boreham, a professor at the University College Dublin Institute for Sport and Health, "is that stair-climbing is one of the few everyday activities at a moderate to high intensity that one can do surreptitiously without having to change, use special equipment or look foolish."

Competitive stair-climbs for charity are a growing phenomenon. The TowerRunning.com website (motto: "Take the stairs and not the elevator") lists well over 100 events around the world, and Italian scientists have analyzed the physics and physiology of these events in a study of "skyscraper running" that appeared in 2010 in the *Scandinavian Journal of Medicine and Science in Sports.* Among the notable insights of the study is that using the handrails to haul yourself up turns the activity into a full-body workout much like rowing, resulting in a "global, maximal effort." About 80 percent of the power you exert goes to raising your body against the force of gravity, 5 percent goes to whipping your limbs back and forth, and the remaining 15 percent goes toward running tiny semicircles at each landing.

Because of its high intensity, stair climbing offers a time-efficient workout: the record for climbing up the Empire State Building is just nine minutes and 33 seconds. However, Boreham and his colleagues have found that a much more moderate approach can also pay dividends. They asked eight undergraduate women to undertake an eight-week program that started with climbing a 199-step staircase twice a day, five days a week. They climbed at a moderate rate of 90 steps a minute, so that it took about two minutes to reach the top. By the end of the program, they were climbing five times a day—not all at once, but scattered through the day—for a daily total of just over 10 minutes of exercise a day.

Compared to a group of matched controls, the stair-climbers increased their aerobic fitness by 17 percent and reduced harmful LDL cholesterol by 8 percent, results that compare favorably to taking a half-hour daily walk. The researchers are now investigating whether the protocol can be transferred to older adults, using a stepping machine rather than staircases. "Because it's at such a high intensity, it accomplishes health adaptations in a shorter period," Boreham says, "which is handy if you like your exercise in short bites."

Standard exercise guidelines suggest that bouts of exercise should last at least 10 minutes in order to produce meaningful gains, although recent research into "high-intensity interval training" (see p. 12) supports the idea that you can get by with short bursts if they're intense enough. The gains in Boreham's study are modest enough that you shouldn't view taking the stairs as the only thing needed to stay fit. But they offer encouraging evidence that simple decisions to be more active in your daily life can add up to measurable health benefits—even if you don't climb the Empire State Building.

CHEAT SHEET: AEROBIC EXERCISE

- No matter what your exercise goals are, aerobic exercise is crucial for your health—and it also plays a key role in sports performance, even in "relaxed" sports like golf.
- Use the "Talk Test" to divide your effort between the aerobic, threshold, and anaerobic effort zones. Spend about 70 percent of your time in the aerobic zone.
- The old "220 minus age" formula for finding your maximum heart rate is highly inaccurate, especially for older adults. (208-0.7 x age) is better, but the only way to get an accurate reading is with a max HR test.
- Altering your breathing to fit a certain pattern or rhythm generally makes you less efficient. If you're panting uncontrollably, you're probably pushing too hard.
- The smooth, unchanging surface of roads and sidewalks may be more of a problem for our legs than hardness. Running on a variety of surfaces minimizes injury risk.
- It is possible, with a lot of hard work, to alter your running form. However, there's no current evidence that doing so will reduce injuries or make you faster.
- Most runners push too hard on uphills and slow down more than they need to on downhills. Practice the mechanics of running downhill to gradually increase your speed.
- The muscles in your arms play very little role in running, but swinging your arms may help keep your legs going through "neural coupling."
- Some riders manage to exceed their "maximum" intensity during spinning classes; researchers believe that motivational instructors and the group setting provide an extra boost.
- Climbing stairs for just two minutes at a time, five times a day, can produce significant fitness gains without even going to the gym.

5: Strength and Power

IN THE EARLY 1900S, ASPIRING WEIGHTLIFTERS visited their local branches of the Institute of Physical Culture, where they attempted to sculpt their bodies into the "perfect" proportions represented in ancient Greek and Roman statues. Among the judges at the first major bodybuilding contest associated with this fad, held at London's Royal Albert Hall in 1901, was Sherlock Holmes creator Sir Arthur Conan Doyle. To the uninitiated, it may appear that not much has changed since then—that the weight room is an intimidating place filled with complicated machines and hulking brutes staring at themselves in wall-sized mirrors.

This is unfortunate, because muscular strength is a crucial ingredient of success in virtually every sport—not to mention its benefits for day-to-day life. And researchers are now realizing that fighting the gradual loss of muscle mass that begins in your 30s is among the most effective tactics to slow down the physical effects of aging. Whether you do it in a weight room or at home, using weights or machines or simply the weight of your body, strength training should be a part of your exercise program. It's no longer just about big muscles—although, if you do want to attain those classical proportions, this is the place to start.

Do I need strength training if I just want to be lean and fit?

If you'll pardon the broad generalization, there are basically two kinds of people at the gym: the ones on the cardio machines hoping to get or stay thin, and the ones on the weight machines trying to bulk up. If you're among the former group, you might think there's not much point in wasting your time doing weights. But you'd be wrong.

For health benefits, aerobic exercise gets most of the attention. It's the go-to activity if you want to reduce your risk of cancer, heart disease, diabetes, dementia, and so on. Strength training, on the other hand, is generally thought of in functional terms: it will help you sprint faster, throw farther, jump higher, and look better in a bathing suit. But there's a lot more overlap than we once thought. For example, recent studies have found that strength training helps people with diabetes regulate their glucose and insulin levels, in some cases more effectively than aerobic training. It can also help control conditions ranging from high blood pressure to depression.

Most important, though, are the benefits you get from strength training that you can't get from other types of exercise. Starting in your 30s, you can expect to lose as much as 1 to 2 percent of your muscle mass each year for the rest of your life. This has incredibly important implications for your prospects of aging gracefully. If you want to be able to move a piece of furniture or lift a bag of groceries—or, later, be able to push yourself up from a chair—you need to hang on to the muscles you've got.

Maintaining your muscle mass also has more subtle effects. The muscle throughout your body is of "high metabolic quality," as McMaster University researcher Stuart Phillips puts it.

That means that it's the primary location for burning fat and taking glucose out of your bloodstream, and the biggest contributor to your resting metabolic rate. As the amount of muscle in your body shrinks, you burn fewer calories and your ability to metabolize food gets worse, leaving you more vulnerable to obesity, diabetes, and other conditions.

Surprisingly, strength training also plays a key role in maintaining strong bones (see p. 177). We tend to think of weight-bearing activity as being the key to bone health—but it's clear that can't be the whole story, otherwise our arm bones would shrink to nothing. Instead, the "mechanostat theory" that emerged in the 1980s argues that the stress that prompts bones to grow stronger is provided by muscles. As a result, strength training is now considered one of the most important ways of keeping your bones strong.

Everybody at the gym has different goals (and of course, they're far more varied than simply losing weight or gaining muscle). But whether you're after health, performance, or simply maintaining the quality of your daily life, you need to spend at least a bit of time on strength training.

How much weight should I lift, and how many times?

The legend of Milo of Croton, a six-time Olympic champion wrestler in the sixth century BC whose training is said to have involved lifting a calf over his head every day until it became a full-grown cow, teaches us two lessons about weight training. First, your workload needs to progress if you want to keep improving. Second, the precise details of what equipment you use and how you use it probably don't matter too much.

As a starting point, the American College of Sports Medicine recommends doing a range of exercises targeting different muscle groups, each with a weight that you're able to lift 8 to 12 times. (You can find the appropriate weight through trial and error over the course of a few workouts and then adjust as you get stronger.) This will produce good results for three or four months—but to maximize your gains after that, you'll need to decide whether you're more interested in developing bigger and stronger muscles, or muscles with greater endurance.

The standard approach to building strength and muscle mass is to emphasize fewer repetitions, lift heavier weights, and take longer rests. A typical workout might be three sets of four to six repetitions for each exercise, taking three minutes of rest between each set. As you become more experienced, you might do a set with anywhere from 1 to 12 repetitions. Varying the amount of weight you're lifting provides a different stimulus, and it's important not to get into a rut where you do the same workout three times a week. (Some important terminology: if you lift a weight 10 times, take a short break, then lift it 10 more times, you've done two "sets" of 10 "repetitions" each.)

Not everyone wants bigger muscles, though, whether for aesthetic or functional reasons. A cyclist, for instance, would rather build the muscular endurance to keep pedaling for hours rather than develop short-lived brute strength. The optimal approach in this case is to do more repetitions using lighter weights and taking less rest. For example, you could do four sets of 20 or more repetitions with less than 90 seconds between sets.

Of course, there are countless different approaches, such as the currently popular CrossFit regimen, which emphasizes short, high-intensity workouts, and the "high-intensity training" system pioneered by Nautilus inventor Arthur Jones. This approach calls for a single set of each exercise, performed at a deliberately slow pace of 15 seconds or more for each lift. Like cow-lifting, these programs undoubtedly produce significant gains if they're done right. Whether they're better than traditional programs is another question.

A 2008 study published in the *Journal of Strength and Conditioning Research* compared traditional strength and muscular endurance regimens in a head-to-head match-up with a "low-velocity" program in which the subjects took 10 seconds to lift each weight and 4 seconds to let it return to the starting position. (The traditional programs took 1 to 2 seconds in both directions.) As expected, the high-weight/low-reps group gained the most strength, and the low-weight/high-reps group gained the most muscular endurance. That left the low-velocity group in the middle. "You can gain [both] strength and muscle endurance," said Sharon Rana, the Ohio University professor who led the study, "but the traditional methods are going to do a slightly better job for those two things."

Similarly, several recent reviews have concluded that single-set training is sufficient for staying fit, but that multiple sets are needed to gain the greatest possible strength. Of course, unless you're focused on a very narrow goal—developing the biggest possible bicep or lifting the heaviest cow—you'll benefit from variety. Keep experimenting with the number of sets, reps, and even lifting speed, and you'll develop a healthy balance of strength and muscular endurance.

How do I tone my muscles without bulking up?

If your upper arm feels a little flabby, you might decide that you need to tone it up. At the gym, you'll do some biceps curls and triceps pulls, with light weights and lots of repetitions, say 30 or more. After all, you don't want to risk adding bulky muscle mass to your arms.

If this sounds familiar, you've fallen prey to a common but mistaken assumption. "Toning," in the sense of light exercise to make a muscle look taut, simply doesn't exist. (In medical terms, muscle tone refers to the low level of tension maintained by the nervous system in all muscles even at rest.) If you're poking somewhere on your body and it feels soft even when the muscle is flexed, that means you're poking fat, not "untoned" muscle. Moreover, the myth of low-weight, high-repetition toning encourages people to use weights so light that they don't have any appreciable effect.

You have two basic options to make your muscles stand out and look more defined: you can make the muscles bigger, or you can shrink the layer of fat covering them. Reducing fat is a whole-body problem—you can't just target the fat on your arms (see Chapter 9 for more on reducing weight). Building muscle has long been thought to depend on lifting weights that are at least 40 to 50 percent of your "one-repetition maximum," the heaviest weight you can lift for that exercise. For advanced resistance trainers, the threshold may be closer to 60 percent. More recent research suggests a simpler rule: whatever weight you use, you should be unable to lift it again when you complete the set. Studies have consistently found that many gym users choose weights that are too low to have any significant effect. For women in particular, avoiding muscle growth is often seen as a good thing. A 2008 study of women working out in New Jersey gyms found that 38 percent of them "believed that the mere act of resistance training would lead to large, 'bulky' muscles." There's an important discussion to be had about gender stereotypes and healthy body images—but even if we accept the unfortunate premise that muscles are bad, the survey still reflects a highly distorted view of what it takes to

CHOOSING THE RIGHT WEIGHT

Strength training guidelines often refer to your "one-rep max," or "1RM"—the heaviest weight that you can lift once for any given exercise. The American College of Sports Medicine suggests that beginners work with weights that are 60 to 70 percent of 1RM for 8 to 12 repetitions; using less than 50 percent of 1RM may not stimulate any muscle growth.

But how do you determine 1RM? For most people, trying to lift the heaviest weight possible is an unnecessary injury risk. Instead, use trial and error to find a weight that has you reaching "failure" near the end of your final set. If you plan three sets of 10 reps and you successfully complete them, increase the weight slightly next time so that you're unable to complete the final one or two reps. New research by Stuart Phillips at McMaster University suggests that reaching failure is the most important factor in building muscle—even more important than how heavy the weights are or how many reps you do. His study found that volunteers lifting at 30 percent 1RM synthesized just as much muscle protein as volunteers lifting at 90 percent 1RM, as long as they lifted to failure.

The take-home: don't worry too much about one-rep max, but choose a weight so that you reach failure, or at least come very close.

build muscle. (Or to put it another way, they should be so lucky!)

Workouts with light weights and a high number of repetitions do have a place in building muscular endurance (see p. 102). Even then, the weight should be heavy enough that you struggle to complete the final set of each exercise, no matter how many repetitions you're doing. But if it's "toning" you're after, your best bet is to choose routines that build muscle size by lifting heavy weights with multiple sets—for example, up to six sets of each exercise, with 6 to 12 repetitions in each set—or else focus on mixing strength, cardio, and diet for overall weight loss.

What's the difference between strength and power?

A major-league fastball takes a little less than half a second to cross home plate. According to a classic 1967 study, the batter has 0.26 to 0.35 seconds to make up his mind, and then 0.19 to 0.28 seconds to swing. To be a great hitter, it doesn't matter if you can bench-press triple your body weight if you can't unleash that strength quickly. *Power* is defined as "force times velocity," and it represents the ability to deliver a large amount of strength in a short period of time. The strength to hoist a heavy weight in a leg-press machine requires only force; the explosive power to leap high in the air requires both force and velocity. That's why power is more important than strength in most sports.

Training for power is subtly different from training for strength. The simplest approach is to lift weights with lighter loads than you normally would, but focus on lifting them with a rapid, explosive motion. The American College of Sports Medicine suggests doing one to three sets of three to six repetitions, using a weight up to 60 percent of the heaviest you can lift for that exercise. But don't just do the same exercises you'd use for strength training; focus instead on functional movements that require multiple joints, like box jumps, jump squats, and medicine-ball tosses. The goal, after all, is to develop an explosive jump, not an explosive hamstring curl.

The specific exercises you choose should be tailored to your goals. For athletes, that means choosing exercises that mimic the motions you'll be using in your sport and doing them at realistic speeds. For example, baseball players are advised to swing the bat at least 100 times a day (three times a week) to increase bat speed, but using a "donut" to make the bat heavier is discouraged because it makes you practice swinging more slowly, according to a 2009 review in the *Journal of Strength and Conditioning*

BUILDING POWER

Training for power is an advanced skill that carries increased risk of injury. Once you've developed your strength and endurance, exercise scientist Greg Wells, the author of *Physical Preparation for Golf*, suggests trying these three power exercises.

SQUAT JUMPS

Stand with feet shoulder width apart. Squat down to a comfortable level, then jump as high as possible. Repeat.

Advanced option: Hold a medicine ball in both hands and throw it as high as possible when you jump.

BOX JUMPS

Stand facing a box or step, feet shoulder width apart. Jump onto the box, aiming for a soft landing. Vary the height of the box as needed.

Advanced option: Start on top of the box and jump down, then rebound by jumping as high as possible into the air as soon as you land.

PLYO PUSH-UPS

Assume the standard push-up position. Lower yourself gently, then explode up and clap your hands before proceeding to the next push-up. Repeat.

Advanced option: Balance your feet on a stability ball.

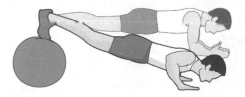

Research. "Explosive" rotational exercises with a medicine ball have also been found to boost power and bat speed.

Even golf, seemingly a much more sedate sport, depends on power. A 2009 study by University of Toronto researcher Greg Wells found that vertical jump—a measure of leg power—was correlated with longer driving distances in elite golfers. It's a lesson that many golfers still haven't learned: "Traditionally, these guys focus on getting stronger and building bulk," Wells says, "but then they slow down and can't hit the ball as hard."

Power isn't just for athletes. When health experts stress the importance of maintaining "functional strength" for day-to-day activities as you get older, they're often talking about power rather than strength. For example, it doesn't require much sustained strength to hoist yourself up from a chair; you need to summon a rapid burst of force to push yourself up. Several recent studies have found that exercise plans incorporating power-building exercises, in which speedy motions with light weights are emphasized, are associated with better outcomes such as improved balance and stronger bones in older adults.

In spite of all that, there can be no power without strength, so don't abandon regular strength training in search of power. But consider incorporating a little explosiveness in your routine, and you may notice the effects on the court and beyond.

Free weights or machines: what's the difference, and which should I use?

The choice between free weights and machines depends on how much stability you want—and that's determined by your goals and your level of experience. For beginners, the biggest advantage of weight machines is that they keep you from making mistakes. Each exercise station is designed to move in only one direction, guiding you through the motion with proper form. But that's also their biggest weakness—because when you call on your muscles in the real world, you won't have that support.

"Weight machines are very stable," says David Behm, an exercise scientist at Memorial University of Newfoundland. "But if you're on a muddy football field or running across a tennis court hitting a forehand on one leg, that's very different." The same is true for everyday challenges, such as getting out of a car for an older person. For that reason, free weights—dumbbells and barbells that aren't connected to pulleys or contraptions—are thought to provide a more functional training stimulus. Because they're less stable, you're forced to balance your entire body while performing the exercise.

Consider a simple exercise such as the biceps curl. If you use a weight machine to perform the curl, you'll strengthen your biceps—which, presumably, is your goal. If, on the other hand, you stand up and do your curls one arm at a time with dumbbells, you'll also be using a host of other muscles such as your back extensors, abdominals, and quadriceps to keep your body upright. With free weights, "people 'cheat' by using other muscle groups," says George Salem, director of the exercise and aging biomechanics research program at the University of Southern California. That "cheating" is a problem if you're using poor form or too much weight and you wrench your back as a result. But it can also be beneficial if you learn to use your whole body to provide stability and added strength, Salem says.

In a quest for even higher levels of instability, some people perform their weights routine while balancing on an inflatable exercise ball. This requires greater effort from the core stability muscles of the trunk and back. But there is a downside, Behm cautions: "To get maximum strength gains, you have

STABILITY FOR STRENGTH TRAINING

The same exercise (in this example, the bench press) can be performed with different levels of stability. More stability is easier and safer for beginners; less stability stimulates more muscles, but limits how much weight you can lift.

MOST BODY STABILITY:
Weight machine

Primary muscles activated:
Pectoral (chest)

LESS BODY STABILITY:
Free weights

Primary muscles activated:
Pectoral (chest)

Secondary muscles activated:
Triceps (back of arm)
Deltoids (shoulders)

LEAST BODY STABILITY:
Free weights with an exercise ball

Primary muscles activated: .
Pectoral (chest)

Secondary muscles activated:
Triceps (back of arm)
Deltoids (shoulders)

Stabilizer muscles activated:
Abdominals

to lift as much weight as possible. But you can't lift as much when you're balancing on a ball." As a result, a 2010 position stand that Behm wrote for the Canadian Society for Exercise Physiology suggests that lifting weights while balanced on an exercise ball is appropriate only for those focused on health and fitness, while athletes seeking performance gains should stick to free weights on a stable surface.

Weight machines do have other advantages, even for experts. They allow you to address specific weaknesses by isolating certain muscle groups, and they're designed to provide a constant resistance through the entire range of motion of each lift. In gyms that stock these machines, the biggest draw may be that they're more time-efficient than fiddling around with free weights. Add it all up and machines are the best choice for many beginners, for both safety and ease of use. Once you've gained some experience, though, it pays to move on, Salem says: "Free weights are more realistic."

Can body-weight exercises like push-ups and sit-ups be as effective as lifting weights?

Joining a health club always seems like a great idea, especially around the beginning of January. That's why 41.5 million Americans pay a total of $18.7 billion a year for health club memberships. But by February many of those new members have discovered that going to the gym can be inconvenient, time-consuming, and sometimes a bit intimidating. On the other hand, buying exercise equipment for the home is an expensive proposition. One solution is to do your workout at home (or in a hotel room or a park or wherever you happen to be), using your own body weight instead of barbells or weight machines for resistance.

There's little doubt that convenience can be a big factor in how well people stick to their exercise programs. A review of studies by the Cochrane Collaboration in 2005 found that patients with conditions like heart disease who were prescribed exercise were more likely to stick with home-based programs than with programs that required them to visit a nearby gym or hospital to use specialized equipment. In the highest-quality study, 68 percent of the home-based exercisers were still doing the program two years later, compared to just 36 percent of the center-based exercisers.

The effectiveness of body-weight exercises depends on your initial fitness and goals. A Japanese study in 2009 tested a body-weight exercise program on a group of volunteers with an average age of 66. The exercise program consisted of leg exercises such as squats, lunges, calf raises, and knee extensions. After 10 months of training twice a week, the volunteers had increased maximum leg force by 15 percent and maximum power by 13 percent. But a closer look at the data revealed that the largest gains were obtained by the subjects who started out weakest, since they had to work hardest to lift their own body weight.

This finding highlights the key weakness of body-weight training: as you get stronger, the weight you're lifting stays the same (or perhaps even decreases, if you're lucky!), which makes it hard to continue progressing. Of course, there are ways to adjust the difficulty of classic exercises like the push-up. You can put your feet up on a chair or use only one arm at a time to increase the difficulty. Adjusting the distance between your hands also changes which muscles are emphasized: if they're closer together than your shoulders, you increase the load on your triceps and deltoids; spreading them farther apart shifts the emphasis to your chest.

If you're a bodybuilder trying to sculpt a Schwarzenegger-esque body, these sorts of adjustments won't be enough to replace the highly specific exercises that you can do at the gym. Similarly, if you're really trying to maximize your strength and power, the weights and machines at the gym allow you to do a wide variety of exercises targeting different muscles while controlling the exact weight, which you simply can't duplicate at home. But for general strengthening, either for fitness or as part of your conditioning for a sport like tennis or basketball, you can get all the challenge you need from a mix of push-ups, pull-ups, crunches, chair dips, squats, and other body-weight exercises. And the price is right!

Can lifting weights fix my lower-back pain?

It's the classic moving-day injury. You're hoisting a dresser or grabbing one end of a sofa, then—bam!—you throw out your back, and your bad lifting technique leaves you unable to straighten up for a week. So it may come as a surprise to hear that a promising solution for chronic lower-back pain, according to a series of recent studies from the University of Alberta, is lifting weights. A whole-body strengthening program dramatically outperforms aerobic exercise for those whose nagging back pain lingers for many months, according to the researchers—and the more you lift, the better.

By some estimates, two-thirds of adults will suffer from lower-back pain at some point in their lives. Many sufferers are diagnosed with "non-specific" back pain, which means their doctor hasn't been able to identify a specific physical problem like a slipped disc or muscle imbalance as the cause. There's no shortage of commonly prescribed solutions, from bed rest and acupuncture to spinal manipulation and radiofrequency denervation, but none have emerged as reliable cure-alls.

Earlier studies have established that not lifting anything neither cures nor prevents this type of back pain. In fact, it can trigger a downward spiral where inactivity makes you weaker, which worsens your back pain and causes you to become even less active, says Robert Kell, a professor at the University of Alberta's Augustana Campus. "People with back pain have a hard time getting through the day, because they get fatigued and are no longer able to maintain their spinal stability," he says. "If you can increase their strength and endurance, they can complete their normal activities without losing their posture."

Since each person's "non-specific" back pain may stem from a slightly different combination of weaknesses and imbalances, Kell uses a 16-week program that targets muscle groups throughout the body. In two studies published in the *Journal of Strength and Conditioning Research,* he has observed more than 25 percent improvement in measures of pain, disability, and quality of life compared with controls and with subjects doing aerobic exercise. A third study, whose results were first presented at the American College of Sports Medicine annual meeting in 2009, divided 240 volunteers into groups who lifted weights between zero and four times a week. Those lifting four days a week decreased pain by 28 percent, compared with 18 percent for three days a week and 14 percent for two days a week. (Non-exercising controls decreased pain by just 2 percent.) "It's much like the exercise recommendations for the general population," Kell says. "If you can make time to do a little bit, like 20 minutes twice a week, it will help. If you can do more, it gets better."

Nonetheless, this one-size-fits-all approach has limitations, according to University of Waterloo professor of spine biomechanics Stuart McGill. "There's actually no such thing as non-specific back pain," he says. "It just means you haven't had an adequate assessment." People with back pain caused by weakness in one of the muscles targeted by Kell's program will indeed see improvement. But the blanket "non-specific" diagnosis also includes people with other sources of back pain, whose condition could worsen if they lift weights. The most prudent course of action, McGill says, is to find a clinician who is able to diagnose the root cause of your back pain and use that diagnosis to determine the appropriate treatment.

Certainly, you shouldn't persist with any exercise that causes pain or discomfort (a rule Kell enforced during the studies). With that caveat, a whole-body strengthening program still seems like an excellent recommendation—whether or not it cures your back, you'll be healthier as a result.

Will I get a better workout if I hire a personal trainer?

In a famous study at Ball State University in Indiana, researchers put two groups of 10 men through identical 12-week strength training programs. The groups were evenly matched when they started, and they did the same combination of exercises, the same number of times, with the same amount of rest. At the end of the experiment, one group had gained 32 percent more upper-body strength and 47 percent more lower-body strength than the other. No performance-enhancing pills were involved—the only difference was that the more successful group had a personal trainer watching over their workouts.

A good personal trainer—certified by an organization such as the National Strength and Conditioning Association, the American College of Sports Medicine, or Can-Fit-Pro—will help you assess your fitness goals, design a safe and effective program to meet those goals, and motivate you to put in the necessary work. But, as the Ball State study shows, there are other, less obvious ingredients that successful trainers provide—and a series of recent studies offers some hints about how we can tap in to these benefits.

The crucial difference between the training of the two groups at Ball State was very simple: by the halfway point of the program, the supervised group was choosing to lift heavier weights. Since both groups started with the same motivation level, it was the trainer's presence leading that group to set more ambitious targets. Other studies have consistently found that, left to their own devices, novice weightlifters tend to work out with weights that are less than 50 percent of their one-repetition maximum, which is too low to maximize gains in strength and muscle size.

Even experienced strength trainers often fall into this trap, according to a 2008 study in the *Journal of Strength and Conditioning Research*. Researchers at the College of New Jersey found that experienced women who trained on their own chose to use an average of just 42 percent of their one-rep max for a 10-repetition set. In contrast, women who had prior experience with personal trainers chose weights averaging 51 percent of one-rep max, even when the trainers weren't there. "Many times, there is initial fear," says Nicholas Ratamess, the study's lead author. "We also found that some women who did not have a personal trainer underestimated their own abilities because they did not routinely push themselves too far."

The latest attempt to address this question comes from researchers at the University of Brasilia in Brazil. They compared two groups of 100 volunteers who undertook a 12-week strength training program, supervised either by one trainer for every five athletes, or one trainer for every 25 athletes. The results display a familiar pattern: the highly supervised group improved their bench press by 16 percent, while the less supervised group chose lighter weights and improved by only 10 percent.

In one sense, this is yet another argument for getting a personal trainer if you can afford one. But the differences here are more subtle, since both groups had access to a trainer who could provide guidance on proper form and choosing appropriate weights. Instead, motivation and the willingness to tackle ambitious goals seem to be the differentiating factors. As Ratamess points out, these are the kinds of benefits that an enthusiastic training partner can also provide. For less experienced exercisers, the educational role of the personal trainer takes on greater importance, he cautions. But beyond that, simply having someone there watching you—whether it's a personal trainer or a workout partner—seems to confer an additional benefit. Certainly, he says, "both have advantages compared to training independently."

Do I need extra protein to build muscle?

It's a pretty safe bet that the guy at the gym who is built like a tree trunk and bench-presses the entire rack also has an enormous barrel of protein powder tucked into his gym bag. This, you might think, is a pretty good endorsement of the "you've got to eat muscle to build muscle" school of thought. But correlation is not the same as causation. "It's hard to argue against years of practice that apparently works," says Stuart Phillips, a McMaster University researcher who studies protein needs in athletes. "The real question is, do they gain muscle because of what they do, or in spite of what they do?"

On this basic question, athletes and scientists remain deeply divided. At McMaster and elsewhere, researchers have spent years conducting careful studies of how much protein exercisers can actually use. By tracking nitrogen, which is found in protein but not in carbohydrates or fat, they can determine whether their subjects are building muscle, losing muscle, or holding steady. "We monitor the food going in and collect the poop, pee, and sweat going out," explains Mark Tarnopolsky, one of Phillips's colleagues. Surprisingly—but consistently—the results show that even serious athletes process only marginally more protein than their sedentary peers, and far less than the megadoses recommended by muscle magazines. Novice weightlifters use the most protein, since they are adding muscle most rapidly, while veteran bodybuilders use less despite their enormous muscles.

That leaves would-be bodybuilders with a choice: Do you take the advice of the egghead in the lab or the muscle-head in the gym? Given that current research techniques aren't perfect, a middle path is likely most appropriate, Phillips says. Although current dietary guidelines in Canada and the United States suggest consuming 0.8 grams of protein per kilogram of bodyweight (g/kg) daily, there is reasonable evidence that 1.1 g/kg is appropriate for serious endurance athletes and 1.3 g/kg for serious strength athletes. (One g/kg is equivalent to about 1.6 ounces of protein for each 100 pounds of body weight.) Even those amounts are below the 1.6 g/kg that average North Americans tend to eat daily when their diet is unrestricted, Tarnopolsky says. That means an ordinary, balanced diet should easily meet your needs—unless you're restricting calories to lose weight. In that case, higher protein intakes (35 percent of calories instead of 15 percent, for example) combined with resistance training appear to help maintain muscle mass while overall mass drops.

Timing also matters: you'll build muscle more effectively if you take in protein within about an hour of finishing your workout. The optimal post-workout dose is about 20 g (0.7 ounces) of protein, according to a 2009 McMaster study in the *American Journal of Clinical Nutrition*. That's equivalent to about 20 ounces of skim milk, four medium eggs, or three ounces of cooked beef. Powders, shakes, and bars offer a convenient way to get this right after a workout—but then again, so does a tuna sandwich.

In the end, if you decide to side with the gym rats and supersize your protein shakes, it's unlikely to do much harm. "The extra protein will, for the record, not pack your kidneys in and will not destroy your bones," Phillips says. The main drawback is that, by taking too much protein, you might end up not getting enough of the carbohydrates that are crucial to performance in both endurance and strength athletes—and that risk should be enough to keep any smart athlete from overdoing the protein.

CHEAT SHEET: STRENGTH AND POWER

- Starting in your 30s, you lose 1 to 2 percent of your muscle mass each year. Strength training can slow this decline and help keep your bones strong.
- A standard beginner's program is one to three sets of 8 to 12 repetitions, reaching failure at the end of the last set. Decrease the number of reps to emphasize maximum strength; increase it to emphasize muscular endurance.
- No matter how much weight you use or how many reps you do, the most important factor in building muscle is reaching muscle failure by the final rep.
- "Toning" muscles with light weights accomplishes little if you're lifting less than 40 to 50 percent of your one-rep maximum.
- Power—the ability to deliver strength in a rapid burst—is more important than absolute strength in many sports. Develop your power by training with explosive movements.
- Weight machines are safe and easy to use, but free weights offer a more "realistic" challenge, forcing you to develop balance and stabilizing muscles.
- A whole-body strengthening program can reduce strain on your back and possibly fix lower-back pain—but you should not persist with any exercises that cause discomfort.
- People exercising under the guidance of a personal trainer gain more strength than those exercising alone, mostly because they're encouraged to lift heavier weights.
- Contrary to conventional wisdom, the amount of protein in a typical North American diet is more than enough to build muscle with any strength training program.

6: Flexibility and Core Strength

To STRETCH OR NOT TO STRETCH? A decade ago, not many people would have dared to even ask that question. But a series of careful studies have upended assumptions that endured for decades, leading to major changes in how scientists view the link between flexibility, injuries, and athletic performance. In particular, it turns out that traditional "static" stretching before exercise not only doesn't help you—it actually reduces your strength, speed, and endurance. Instead, new research suggests that the best way to warm up involves "dynamic" stretching.

Beyond traditional stretching, many people turn to activities like yoga and Pilates to enhance flexibility and build core strength. Researchers are only now beginning to study these activities and draw conclusions about what benefits they can and can't offer.

Will stretching help me avoid injuries?
If there's one topic that perfectly illustrates the divide between sports scientists and athletes of all levels, it's stretching. From peewee to the pros, almost everyone does it—despite the publication of study after study suggesting that stretching won't prevent injuries, won't prevent soreness, and in some circumstances will make you slower and weaker. Hockey players, for example,

are "pathologically obsessed with stretching their hamstrings and groin," says Mike Bracko of the Calgary-based Institute for Hockey Research. The pre-game ritual is so ingrained that the players don't really care what the evidence says.

The evidence, to be fair, is still full of contradictions and unanswered questions even after decades of study. This is partly because there are so many different ways to stretch and because people expect so many different benefits. For clarity, we'll first examine the general question of whether a stretching program can help prevent injury. Then we'll take a separate look at whether pre-exercise stretching helps you produce your best performance (see p. 124), and whether post-exercise stretching helps you recover and avoid soreness (see p. 131).

It seems intuitively obvious that you're most likely to pull a muscle if it's tight. That's why we stretch. The simplest and most common form of stretching is "static" stretching, which involves holding a position at the very edge of your range of motion for, say, 30 seconds at a time. There's no dispute that this form of stretching increases your range of motion, both in the short term (during the subsequent workout) and in the long term (producing lasting changes).

The weak link in the chain of logic is the assumption that being more flexible protects you from injury. Most muscle injuries occur within the normal range of movement during "eccentric" contractions (while the muscle is lengthening, for instance when you're lowering a biceps curl). In other words, you're most likely to pull your hamstring while sprinting or changing direction, not while trying to do the splits (unless you're a ballerina or a hockey goaltender, in which case static stretching is in fact important). "If injuries usually occur within the normal range of motion," McGill University sports doctor Ian Shrier asked in a widely cited editorial in the *British Journal of Sports Medicine* in 2000, "why would an increased range of motion prevent injury?"

Literally hundreds of studies have tried to answer that question, and the Centers for Disease Control and Prevention reviewed 361 of them in 2004 in search of an answer. "Stretching was not significantly associated with a reduction in total injuries," they concluded. Moreover, "use of stretching as a prevention tool against sports injury has been based on intuition and unsystematic observation rather than scientific evidence." Subsequent reviews, most recently in 2008, have reached the same conclusion. One example of the "unsystematic observation" used to justify stretching is that several influential early studies compared one group that warmed up and stretched to another group that did neither. The problem with this study design is that warming up—for example with a gentle jog and some movement drills within the expected range of motion (see p. 128)—reduces injury risk whether you stretch or not.

It's important to note that the lack of proof that stretching prevents injury isn't the same as proving that it doesn't work. It may be that stretching programs have to be tailored to the individual needs of each person or each activity, so studies of a generic stretching program are doomed to produce ambiguous results. Imagine testing your prescription eyeglasses on 1,000 random people and using the results to judge whether eyeglasses help people see more clearly. So if, like Bracko's hockey players, you're deeply attached to your stretching routine, there's not enough evidence to compel you to give it up—but there are some compelling reasons to think very carefully about when and how you stretch.

Could stretching before exercise make me slower and weaker?

Until the debate about whether stretching prevents injuries is settled, many people will keep it as part of their exercise routine—so it's worth considering whether stretching has any other effects that might influence when you choose to stretch. One major change that has already taken hold is the realization that you shouldn't stretch a "cold" muscle. Experiments with leg muscles from rats at the University of Michigan, for example, have shown that even mild stretching is enough to damage unprepared muscle fibers. Since stretching is typically incorporated as part of a pre-exercise warm-up routine, experts now recommend that you first start with a gentle jog (or bike or swim) to warm up your muscles and make them more compliant. You should also avoid overstretching to the point of pain.

Even if you follow this advice, a series of studies in the last few years suggests that you'll still suffer from temporary after-effects that make you weaker, slower, and less efficient. This phenomenon isn't yet fully understood, but some experiments have found that it can last up to two hours after stretching—not exactly the ideal preparation for a workout or competition where you want to be at your best. Consider, for example, a 2010 study by researchers at the University of Milan, published in the *Journal of Strength and Conditioning Research*. A group of 17 volunteers performed a series of vertical jumps from different squatting positions, with or without stretching their leg muscles beforehand. The height of the jumps, the peak force they exerted with their legs, and the maximum velocity attained were all significantly lower after stretching, in agreement with similar results from a half-dozen earlier studies.

Why does this happen? There are various theories—for example, it's possible that "looser" muscles and tendons aren't able to transmit force to the bones as effectively, much like the ropes on a sailboat need to be taut in order to trim the sails. Or, on a microscopic level, it may be that individual muscle fibers are able to exert more force when they're shorter. There is also evidence for "neuromuscular" effects, in which the after-effects of stretching disrupt the signal between brain and muscle calling for a contraction. Most likely, it's a combination of these factors—that would explain why researchers at McMaster University found decreased strength related to neural signaling that lasted for about 15 minutes, along with weakness in the muscles themselves that lasted for up to an hour after a session of calf muscle stretching.

Of course, observations in the laboratory don't always translate to the real world. So scientists and coaches at Louisiana State University's NCAA-champion track team put the theory to the test with 19 of their star sprinters in 2008. Each athlete performed three 40-meter sprints in two different sessions separated by a week; they performed a dynamic warm-up (see p. 128) before both sessions and added four static stretches of the calf and thigh muscles before one of the sessions. The result: they were significantly slower (5.62 versus 5.72 seconds) when they stretched, with most of the deficit coming in the second half of the sprint.

All of this evidence, along with related studies that show endurance may also be compromised, makes a compelling case that you should avoid stretching before workouts or competitions, says Jason Winchester, the lead author of the LSU study, whose consulting clients include National Football League teams and the U.S. national track team. Instead, incorporate light stretching after your workouts or on rest days. If you really prefer stretching before workouts, the slight loss of strength and speed is unlikely to make much difference, but if it's a competition, leave as much time as possible between stretching and the start.

Do flexible runners run more efficiently?

Even if we can't agree on what the "perfect" running form is (see p. 88), most of us do recognize when we see a runner who looks particularly smooth and effortless. Seeing it is one thing, but imitating it is another thing entirely—no one is quite sure how to develop a more efficient running style. One of the most common ways people try to do this is to increase flexibility with a regular stretching program, so that they won't be held back by tight hamstrings or locked-up hips. This makes sense intuitively, but the evidence doesn't back it up.

When researchers tinker with running form, they're less interested in subjective outcomes like how smooth you look than in more objective measures like "running economy"—a concept similar to fuel economy in cars. Running economy tells you how much energy you have to burn to run at a given pace; the less energy you burn, the longer you'll be able to continue at that pace. (It's usually determined by measuring very precisely how much oxygen you breathe in and how much is successfully transported to your working muscles.)

Researchers have been sticking volunteers on treadmills for decades, trying to figure out which factors lead to good or bad running economy. As far back as 1990, studies were starting to suggest that the runners with the most trunk and lower-body flexibility had the worst running economy. A rigorous study of 34 world-class British distance runners in 2002 reached the same conclusion, using a simple sit-and-reach test, which involves sitting on the ground with your legs straight in front of you and reaching as far as possible toward your toes. Another study, at Nebraska Wesleyan University in 2009, found the same connection between better sit-and-reach score and worse running economy.

This effect likely stems from the remarkable ability of your muscles and tendons to store energy like coiled springs, providing an estimated 40 to 50 percent of the energy you use for each step. "If you decrease the stiffness of the muscles and tendons, then you can't store and reutilize energy as well," explains Jacob Wilson, an exercise physiologist at Florida State University. The connection is less pronounced in women, who are generally more flexible to start with. And the initial studies only measured correlations, rather than showing that increasing your flexibility actually causes lower running economy.

To address this gap, Wilson and his colleagues asked 10 male runners to complete a pair of one-hour tests consisting of 30 minutes running at a predetermined pace to measure running economy, followed by 30 minutes as fast as possible. Before one of the tests, the subjects did 16 minutes of "static" stretching—the most common technique, which involves stretching a muscle to the edge of its range of motion, then holding for 30 seconds. Sure enough, the non-stretchers burned about 5 percent fewer calories in the first part of the run and ran 3.4 percent farther in the second part. Though it was already well established that static stretching causes a temporary decrease in strength and power, the 2010 study marked the first time the effect had been observed in an endurance activity.

That doesn't mean you should leap into exercising with no warm-up—it just means you should rethink how you warm up. New results from a follow-up study by Wilson and his colleagues suggest that, unlike traditional static stretching, an alternative approach called "dynamic" stretching doesn't hurt running performance.

How should I warm up before exercise?

After reading the last few pages about the negative effects of static stretching before your workout, you might be thinking, "Great! Now, instead of wasting my time warming up, I can get down to business right away." Nothing could be further from the truth. In fact, the stretching studies reinforce the point that subtle differences in how you prepare your body can make a big difference in your performance during a workout or competition.

The general goals of a warm-up are "to increase muscle and tendon suppleness, to stimulate blood flow to the periphery, to increase body temperature, and to enhance free, coordinated movement," according to a group of U.S. Army researchers who studied the problem in 2006. A gentle jog accomplishes some of these goals—raising body temperature, for instance—but it doesn't do much to prepare the specific muscles that will help you lift a weight, throw a ball, or cut sideways across the court. Instead, you need to perform a series of exercises that move your muscles through the full range of motion that you plan to use, at first gently and then with increasing vigor. This is a dynamic warm-up, focused on movement rather than the static poses of a traditional stretching routine.

Over the past decade, a series of studies has tested the principles of dynamic warm-up. The U.S. Army study had recruits perform one of two 10-minute warm-up routines, one dynamic and one static. Those who performed the dynamic warm-up produced significantly better performances in three tests of agility and power (a shuttle run, an underhand medicine-ball throw, and a five-step jump), compared with static stretchers and those who did no warm-up at all. Other dynamic warm-up studies have found improvements in vertical jump, bicycle sprint, oxygen uptake, and even coordination.

Most of these studies focus on the acute effects of warming up—after all, we're most interested in how our warm-up affects the workout that follows. But University of Wyoming researchers posed an interesting question in a 2008 study: What are the long-term benefits of repeated dynamic warm-ups? Using the same warm-up routine as the Army study, the researchers monitored a group of collegiate wrestlers for four weeks. At the end of the trial, the dynamic warm-up group had improved on a whole battery of tests of strength, endurance, agility, and anaerobic capacity (for example, broad jump, sit-ups, push-ups, 600-meter run, and so on). A matched group that did static stretching instead of the dynamic warm-up saw no improvement in any of the tests.

The precise details of a dynamic warm-up depend on the demands of the activity you're preparing for, but Louisiana State University researcher Jason Winchester suggests breaking it down into three basic stages:

1. A low-intensity, rhythmic activity to elevate heart rate and body temperature; for example, at least five minutes of jogging, swimming, or biking.
2. A few minutes of dynamic drills that put your muscles through the range of motion you'll be using; for example, squats, arm windmills, and skipping. Do 10 repetitions of each one.

DYNAMIC STRETCHING

To warm up before a race or workout, start with five minutes of gentle jogging. Then do 10 repetitions of each of the following drills:

HIGH KNEES
Run slowly forward with an exaggerated knee lift. Take very short steps and don't lean back.

HEEL KICKS
Run slowly forward while kicking your heel up to touch your buttocks. Take short steps, mainly moving your lower legs.

WALKING LUNGE
Take a long, exaggerated step, letting your body sink until your front thigh is parallel to the ground. Then take a step with the opposite leg and repeat.

3. Finish with some skill-specific motions to prepare for your activity. If you're lifting weights, lift a few reps with a light load; before a tennis match, hit some easy ground strokes; or run a few relaxed sprints before a hard run.

These basic principles can be adapted for just about any sport or physical activity, focusing on motions that prepare your muscles for the challenges ahead. The more vigorous or explosive the activity, the more thoroughly you should warm up.

Will stretching after exercise help me avoid next-day soreness?

In 1986, researchers at the Free University of Amsterdam asked a group of volunteers to perform a set of strenuous exercises with one leg, while resting the other leg. Over the next three days, the scientists poked, prodded, and measured the legs in an effort to understand why certain exercises cause us to feel sore—not immediately after the exercise, but usually beginning the next day and often peaking two days later. One of the tests involved using electrodes to record the electrical activity in each leg, in order to look for differences between the sore leg and the rested leg. None were found. This was a very significant result, because it helped rule out a dominant theory of "delayed-onset muscle soreness," or DOMS (see p. 61), that was first proposed in the 1960s. The theory argued that, after heavy exercise, damaged muscles went into spasm, blocking blood flow and causing the observed pain. The Dutch experiment put an end to that idea.

As often happens, the advice spawned by a discredited theory lived on long after the theory. Researchers in the 1960s had proposed that the best way to deal with these hypothetical spasms was to stretch the affected muscle, allowing normal blood flow to resume. Exercisers obediently began stretching after exercise in the hope of avoiding next-day soreness. And to this day, they still do.

Given that the spasm theory has been abandoned, it shouldn't come as much of a surprise that studies of post-exercise stretching have found little or no effect on soreness. The most recent lab experiment, published in 2009, put a group of 20 Australian rowers through a series of grueling stair-climbing workouts (the muscle contractions required to go down stairs or down hills are particularly effective in causing soreness). Some of the rowers did a 15-minute static stretching routine after the workout, while others simply rested for 15 minutes; a week later, they switched. Over the three days following each workout, the researchers found no differences at all between the two groups in muscle strength, perceived soreness, or blood levels of a marker of muscle damage called creatine kinase.

Other researchers have tried similar experiments outside the lab. For instance, an Australian football team allocated different players to different recovery protocols, including either rest or a 15-minute post-game stretching session, each week over the course of 12 weeks. On a whole host of measures, including soreness, vertical jump, peak power on a stationary bike, and flexibility, there was no difference between the groups.

The trend in these studies is consistent. An independent review of 25 studies on the topic, published in 2008 by the Cochrane Collaboration, found "very consistent" evidence that stretching has "minimal or no effect on the muscle soreness experienced between half a day and three days after [exercise]." Of course, you may have other reasons to stretch after a workout. If you're hoping to increase flexibility, that's the best time to stretch, since your muscles are still warm and the performance-dampening effects of stretching don't matter once the workout's over. But, sadly, it won't stop you from getting sore.

Where is my "core," and do I need to strengthen it?

These days, it's all about the core. Whether it's yoga, Pilates, exercise balls, or dozens of other fitness programs and gadgets, there's no greater selling point than a promise to improve your core stability. And researchers now agree—for the most part—that weak core muscles can indeed be a key culprit in everything from lower back pain to sports injuries, says Reed Ferber, a kinesiology professor who directs the University of Calgary's Running Injury Clinic. "What people don't agree on," he adds, "is what the core is."

In particular, there's a tendency to focus too much on the abdominal and lower back muscles. But pelvic and hip muscles also play a crucial role in stabilizing the body during activity and are now generally considered part of the "core." Ferber cites the example of a 40-year-old woman who came to him as a patient with knee pain. "She had a fantastic six-pack, and did Pilates or yoga six days a week," he says. But the woman was unable to balance long enough to do a simple one-legged squat—bending at the knee while standing on one leg—because her hip muscles weren't strong enough to provide balance. This instability was the root cause of her knee injury.

The same pattern was borne out by a seven-month study of patients at Ferber's clinic, 92 percent of whom turned out to have abnormally weak hip muscles (and 89 percent of whom improved with four to six weeks of hip strengthening). Similarly, a University of Delaware study of basketball and track athletes found that the best predictor of who would develop leg injuries during the season was weakness in one of the hip muscles.

Even for the abdominal muscles, not all exercises are created equal. A study presented at the American College of Sports Medicine's 2008 annual meeting found that traditional crunches, which involve curling the torso up, mainly activate superficial "six-pack" muscles rather than the "deep abdominal" muscles that are more crucial for stability. The study, by Auburn University researcher Michele Olson, used EMG electrodes to compare muscle activation produced by various core exercises. Pilates exercises in which the torso stays unflexed, like the "Hundred" (lying on your back and lifting your legs at a 45-degree angle, with your arms at your side) and the "Double Leg Stretch" (similar to the "Hundred" but with your arms also raised at 45 degrees behind your head), were more effective than crunches at strengthening the deep abs instead of superficial muscles.

For elite athletes, designing a core program often begins with a detailed assessment of their strengths and weaknesses, in order to target the weakest areas. But that remains more of an art than a science, Ferber says, so most people would benefit from a fairly general core program—one that includes an activity like Pilates, but also incorporates some more functional exercises that mimic the range of motion used in whatever activity they participate in. His top suggestion, based on the people he sees at his clinic, is hip exercises, which are relevant for sports ranging from soccer to cycling. And working on the hips offers an important reminder, he adds: "The bottom line is a six-pack does not equal core stability."

HIP STRENGTHENING

Pelvic and hip muscles play a crucial role in stabilizing the body during activity, so they're part of the body's "core" and should be exercised along with the abdominal and lower-back muscles.

Try these exercises after (not before) workouts, gradually building up to three sets of 10 repetitions. For each of these motions, count two seconds in and two seconds out, controlling the motion throughout.

HIP ABDUCTOR - STANDING
1. Place opposite foot behind the leg using resistance band.
2. Move band leg outward, keeping knee straight.

HIP FLEXOR - STANDING
1. Place opposite foot beside leg with band.
2. Move involved leg forward, keeping knee straight or with slight "soft knee."

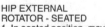

HIP EXTERNAL ROTATOR - SEATED
1. In seated position, move leg outward and return to starting position slowly.
2. Keep knees together.

What are the benefits of yoga for physical fitness?

A few years ago, researchers at the University of Nevada, Las Vegas, found that a brief yoga session allowed high school track athletes to improve their time for a one-mile run by an average margin of one second. Not a particularly earth-shattering result—especially when you realize the same researchers also found that 20 minutes of motivational shouting ("You're the definition of speed!") improved performance by five seconds.

The point here is that it's very difficult to tease apart the mental and physical components of athletic performance. That's especially true for yoga, a word whose very meaning— "to yoke, or unite"—refers to the goal of integrating body, breath, and mind. "Yoga would say that there's really no way to affect the mind without affecting the body, and there's no way to affect the body without affecting the mind," says Timothy McCall, a San Francisco doctor and author of the 2007 book *Yoga as Medicine*. Increasingly, though, people are turning to yoga in search of physical benefits. According to a 2008 survey, 15.8 million Americans were practicing yoga, and 49.4 percent of them started with the goal of improving their health (compared to just 5.6 percent in a similar survey five years earlier).

Considered strictly as a form of exercise, yoga has many strengths and a few key weaknesses. A 2001 study at the University of California, Davis, found that an eight-week program of two 90-minute hatha yoga classes per week led to significant increases in strength, muscular endurance, and flexibility. Other studies have found improvements in balance and even bone density. This is not really surprising—yoga is, after all, a weight-bearing physical activity. If you want to know how much you'll improve, that's a more difficult question, because there's such a bewildering array of yoga styles. Like any other form of exercise, the benefits depend on which specific activities you do, how often you do them, and how vigorously.

A more controversial question is whether yoga is vigorous enough to count as aerobic exercise that improves cardiovascular fitness. While the UC-Davis study observed a small but measurable improvement of 6 percent in aerobic fitness, other studies have failed to confirm this. In a 2002 Northern Illinois University study of 45-minute "power" (ashtanga vinyasa) yoga sessions, the participants' heart rates stayed below the recommended threshold for aerobic exercise. Similarly, a 2007 study found that average energy expenditure during a hatha yoga session was equivalent only to a leisurely stroll, although heart rates did reach a moderate level during the "sun salutations" portion of class.

That doesn't mean yoga can't provide a good aerobic stimulus. Karen Rzesutko, the lead researcher in the Northern Illinois study, says she believes an experienced power yoga practitioner "who is motivated enough to give 100 per cent to the practice" can reach and maintain a suitably high heart rate. But at a certain point, it's probably best to accept yoga's strengths and weaknesses rather than fight them, and seek your aerobic exercise in other ways. After all, McCall points out, having a balanced approach is a worthy yogic principle—that's why he hikes, bikes, and dances in addition to doing yoga.

What are the benefits of yoga for overall wellness?

Of the eight "limbs" of classical yoga, formulated over 2,000 years ago, only one relates to physical fitness as we now think of it. The others include ethical principles, the flow of vital energy, and meditation as paths toward self-knowledge and enlightenment. These days, many people have a more casual attitude to yoga. But even if you just drop in to an occasional class, you'll spend time on breathing and concentration exercises that you wouldn't encounter in a typical gym routine—part of the discipline's continuing commitment to a state of wellness defined much more broadly than the usual fitness markers.

Science, needless to say, seeks to classify and measure these benefits. One theory is that yoga helps to control the "fight-or-flight" response that physical, mental, or emotional stress triggers in your body. The stress hormone cortisol, for instance, triggers a cascade of physiologic, behavioral, and psychological effects through the endocrine system; similarly, your nervous system responds to stress by sending signals to elevate heart rate, blood viscosity, and blood pressure. If these responses are triggered too often, your body gets run down and susceptible to illness.

Several studies have found that yoga programs can lower levels of cortisol throughout the day, including a 2009 randomized trial that compared yoga with "supportive therapy" in a group of 88 breast cancer patients. (In contrast, yoga failed to change cortisol levels in another 2009 study of women with rheumatoid arthritis.) Various other studies have seen positive effects from yoga on outcomes like perceived stress, mood, and sleep patterns.

Of course, researchers have observed many of the same benefits from non-yogic exercise. A Rutgers University study in 2008 pitted hatha yoga against strength training in a head-to-head comparison, putting subjects through 50-minute sessions and then measuring the effects on anxiety, tension, calmness, and other mental health variables at 15-minute intervals afterwards. Both yoga and strength training had positive effects— yoga improved scores on anxiety and calmness, while resistance training improved all variables. Interestingly, yoga's impact was most pronounced immediately afterwards, then began to fade within an hour. Strength training, in contrast, produced effects that intensified as recovery progressed, suggesting a longer-lasting result.

It's important to note that the weights session was perceived by participants to be "moderate exercise," while the yoga session was "light exercise"—a difference that could explain why weights had a bigger effect. But, lead author Joseph Pellegrino explains, the details of the sessions were carefully chosen to mimic how people really do weights and yoga, in order to do a real-world comparison.

Other studies have found, in general, more similarities than differences between yoga and other forms of exercise. A review of 81 studies by University of Maryland researchers in 2010 concluded that "yoga may be as effective as or better than exercise" for a variety of health-related outcomes but acknowledged a lack of rigorous studies. For now, researchers haven't been able to isolate and identify any secret ingredients that yoga offers and other forms of exercise don't. But it's clear that, whether you choose a yoga class or a relaxing bike ride along a waterfront path, you'll be getting benefits for both body and mind.

CHEAT SHEET: FLEXIBILITY AND CORE STRENGTH

- Stretching increases your range of motion, but studies have failed to confirm that stretching reduces injuries. The best time to stretch for flexibility is after exercise, not before.
- "Static" stretching reduces strength, power, and speed for an hour or more, thanks to a combination of neuromuscular effects and lowered force transmission in "loose" muscles and tendons.
- Runners who display greater flexibility in a sit-and-reach test run less efficiently, and pre-run static stretching also lowers efficiency and worsens performance.
- Warming up with "dynamic" stretching exercises raises the temperature of muscles and prepares them for exertion but doesn't decrease strength, power, speed, or endurance.
- Stretching after exercise makes no difference to how sore you are the next day.
- Hip muscles and deep abdominal muscles are more important than the superficial "six-pack" muscles for core stability and injury prevention.
- The benefits of yoga depend on the style and level; in general, yoga classes are good for flexibility and strength but are insufficient to count as an aerobic workout.
- Like other forms of exercise, yoga can help reduce stress hormones and control mood.

7: Injuries and Recovery

When Greek physician Herodikos of Selymbria, sometimes considered the father of sports medicine, got tuberculosis, he treated it with a vigorous program of massage, steam baths, and wrestling. We've come a long way in the 2,500 years since then (wrestling is out, massages are still in, and steam baths . . . well, it depends), but one principle hasn't changed since Herodikos's time: it's better to prevent an injury, or at least nip it in the bud, than to treat it once it's already full-blown.

Not surprisingly, contact sports like hockey and rugby cause the most injuries among adolescent boys (soccer and basketball lead the way for adolescent girls). But non-contact sports are capable of producing overuse conditions like tennis elbow and runner's knee. By some estimates, in fact, a staggering 70 percent of recreational runners get an injury in any given year. With that in mind, it's important to understand that occasional injuries are an inescapable part of exercise—but with the right care, you can make a rapid return to full strength.

Ouch, I think I sprained something. How long should I stay off it?

When figure skater Anabelle Langlois fractured her fibula in a training accident just over a year before the 2010 Olympics, doctors pursued every possible avenue for her rehabilitation, including two surgeries. One thing they didn't recommend, though, was a long period of complete rest for the injured leg.

In the past few decades, doctors have changed their thinking about the best treatment for sports injuries, ranging from sprained ankles and pulled muscles to, in some cases, broken bones. After the acute pain and swelling has passed—sometimes in as little as a few days—movement and gentle loading of the injured area seems to help muscles heal better, hasten return to full strength, and reduce the risk of recurrence. That advice remains little heeded, in part because of the very real risks of pushing too soon, and in part because of our natural caution. "In your head, you want to protect an injury," says Langlois. But at the urging of her doctor, she was putting weight on her injured leg within two weeks of being operated on, with a still-broken bone and an open surgical scar. "That really surprised me," she says.

The goal of early mobilization isn't just to return an injured athlete to competition as quickly as possible. Favoring an injury for too long causes muscles to atrophy from disuse and affects the healing process. "If an injured muscle heals without any stress being put on it, it will generally heal in a shortened position, and the affected area will be a bit weaker and more fibrotic than the surrounding tissue," says Shawn Thistle, a lecturer in the orthopedics department of the Canadian Memorial Chiropractic College. "It ends up being the weak link when you return to activity."

A study published last year in *Histology and Histopathology* illustrates the process. Brazilian researchers compared the recovery of rats that rested against those that began moving their legs either an hour or three days after a muscle injury. Both mobilized groups were able to regenerate more muscle fibers than the rested controls, but only the early mobilization group also had a decrease in fibrotic scar tissue. Humans and rats recover at different rates, so it's impossible to apply these findings directly to humans (and it's equally difficult to find a group of human subjects with identical injuries to conduct a similar experiment), but the general principle of early mobilization is the same.

There are limits, though: before you start mobilizing the injured muscle, you have to give it a chance to form scar tissue strong enough to prevent re-tearing. During this period, which for mild injuries may last three to seven days, the "RICE" protocol of rest, ice, compression, and elevation helps speed recovery. Once the acute phase has passed, activity can progress in a sequence beginning with simply moving the affected muscle through its range of motion, then load-bearing exercises, and eventually functional activities, says Thistle. You can think of it as "MICE" rather than "RICE," where movement replaces rest. Pain can serve as a useful guide to tell you when you're pushing farther than you should.

Of course, elite athletes aside, most people won't have a team of physicians carefully monitoring their progress, which makes it risky to push the pace of rehab too much. Any injury in which the initial pain and swelling persist for more than a day or two should be evaluated by a doctor or sports therapist. But for the milder tweaks that inevitably accompany many sports, it's worth bearing the principle of "active rehab" in mind. Reestablish the full range of motion as soon as possible, and follow up by loading the muscle. Don't push to the point of pain, but don't hobble yourself by protecting an injury long after it's healed, either.

ACTIVE REHABILITATION: SPRAINED ANKLE

It used to be that sprained ankles, which are among the most common sports injuries, were completely immobilized in a plaster cast. These days, physicians and therapists emphasize functional rehabilitation, which proceeds in four basic stages.

1. RANGE OF MOTION
For mild ankle sprains, the process of regaining range of motion can begin immediately after injury.

EXERCISE:
ACHILLES TENDON STRETCH
Use a towel to pull your foot toward your face, stopping if you feel pain. Do five repetitions, holding for 15 to 30 seconds.

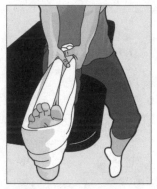

2. PROGRESSIVE MUSCLE STRENGTHENING
Once swelling and pain subside, try simple static exercises like pushing your foot down against the floor. Progress to exercises requiring movement.

EXERCISE: ANKLE EVERSION
Using a rubber band for resistance, rotate the foot away from the center of the body. Count one second as you push the foot out, and four seconds as you let it return.

3. PROPRIOCEPTIVE TRAINING
When the ankle can bear weight without pain, begin recovering balance and postural control to avoid future sprains.

EXERCISE: WOBBLE BOARD
While seated, place your foot on the board and rotate it clockwise and counterclockwise without lifting your foot. Progress to standing up.

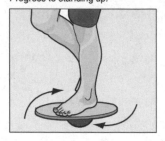

4. ACTIVITY-SPECIFIC TRAINING
Once you can walk without pain, start preparing for the particular sports and activities you hope to resume.

EXERCISE: FIGURE-EIGHTS
Start with a mix of walking and running, then try more difficult tasks like backwards running or running in patterns like circles and figure-eights.

Will a post-exercise ice bath help me recover more quickly?

As appealing as the prospect of a soak in the hot tub after a workout may sound, the consensus among elite athletes is that you're better off doing the opposite. Ice baths at a chilly 50 to 59°F (10 to 15°C) have become the first line of defense for athletes dealing with everything from the heavy impact of bone-jarring tackles to the repetitive stress of a marathoner's three-hour run. Two-time Olympic miler Kevin Sullivan, for example, soaks his legs in the cold tub several times a week, immediately after hard workouts. "And then if I'm feeling a little stiff or tired heading into a race," he adds, "I'll try to use them in the days immediately leading up to the race."

The logic behind ice baths relates to the normal wear and tear of exercise. Hard exertion causes "microtears" in your muscles; these microtears stimulate new growth that makes you stronger, says University of Toronto exercise physiologist Greg Wells. But this damage can also cause soreness that interferes with the next day's workout, so quick repair is essential. Ice baths cause blood vessels to constrict, forcing waste products out of the affected area. "It's almost like wringing out a sponge," Wells says. Then, when the area warms up again, fresh blood rushes in to help the healing process.

At least, that's the theory. But scientists putting ice baths to the test under laboratory conditions have produced mixed results. One problem is that different studies have used different protocols for their cold baths, making it hard to compare. For example:

- A 2007 study in the *British Journal of Sports Medicine* tried using three one-minute dunks in 41°F (5°C) water, with

one minute between dunks. They found no benefits in
perceived soreness, swelling, or blood markers of muscle
damage after a leg workout compared to room-temperature
water or no bath at all.

- A 2009 study of Australian soccer players tried five one-
minute immersions in 50°F (10°C) water, again with one
minute off between dunks. This time, the soccer players
felt less sore and less tired a day later (compared to a body-
temperature bath), but there was no change in how quickly
their strength returned or in blood markers of damage.

- Another Australian study in 2009 tried two five-minute
sessions in 50°F water, separated by a 2.5-minute break.
This time, they observed decreased soreness and quicker
recovery of strength and speed, though the changes still
weren't reflected in blood markers of muscle damage.

The latter study, by researchers at the University of Western
Australia, also compared their protocol with the results from
another popular technique called "contrast therapy," which involved six alternating two-minute bouts of cold (41°F) water
with warm (104°F/40°C) water. The goal of contrast therapy is
to "squeeze the sponge" several times instead of just once, but
the results were significantly worse than cold alone. The problem may be that the two-minute bouts don't allow enough time
for the deep muscle tissue to actually change temperature, the
researchers suggest—which may also explain why the ice-bath
tests that dunk their subjects for only one minute at a time have
produced disappointing results.

Despite the difficulties in determining exactly how and why
ice baths work, most researchers—even University of Melbourne
professor Peter Brukner, one of the authors of the *British Journal*

of Sports Medicine study that found no benefits—are cautiously optimistic that the baths offer real therapeutic value. "Even though our research was unconvincing, I still encourage their use," Brukner says.

You can make your own ice bath in the bathtub or in a (clean) garbage bin, with a few trays of ice or a reusable ice pack. Better yet, take advantage of a cold river or lake to soak your legs after a hard workout. Given the studies described above, you should aim for 5 to 10 minutes to allow the cooling effect to penetrate. (Keeping a pair of socks on can make your feet more comfortable; don't soak in temperatures below 41°F, as there's a risk of tissue damage.)

Will a heat pack or hot bath soothe my aching body?

Nowhere are the healing powers of a hot bath more respected than in Japan. Researchers there have found, for example, that levels of the stress hormone cortisol drop after a relaxing soak in the tub. Heat is also prescribed for conditions ranging from arthritis to chronic pain. But when it comes to exercise and athletic injuries, heat isn't always the right choice.

The traditional advice has been to apply cold to acute injuries—a sprain or a bruise, for example—and reserve heat for nagging pains that persist for weeks or months. The reason is that new injuries are often accompanied by swelling. Cold constricts your blood vessels to limit swelling, while heat can have the opposite—undesirable—effect. Chronic injuries, on the other hand, are often tight and surrounded by scar tissue, so heat can help soften and loosen the muscles around the injury, allowing them to move more freely.

Researchers believe that to have a significant loosening effect on muscle, the heat needs to increase skin temperature by

5 to 7°F (3 to 4°C) for about five minutes. The problem is that when you put a heat pack next to your skin, the greatest heating effect is limited to the outer quarter-inch of your body. Even just an inch below the surface, a heat pack generally elevates muscle temperature by less than 2°F. Electric blankets, hot water bottles, saunas, and even hot baths also qualify as "superficial" heat sources that don't penetrate far into your muscles. (Heating deeper tissue generally requires machines using, for example, ultrasound, shortwave, or microwave energy. Several studies have found that shortwave machines can produce muscle temperature changes of over 7°F at a depth of greater than an inch.)

So does heat work? According to a 2010 literature review by the Cochrane Collaboration, there is "moderate evidence in a small number of trials" that heat wraps can reduce lower-back pain. For example, a pair of studies found that after five days of using a heat wrap, subjects reduced their back pain by 17 percent compared to subjects who were given a placebo pill instead. This gives some credence to the idea that you might get some relief from nagging aches throughout the body by applying a heat pack, or even by soaking in the tub.

But the more common and well-supported use for heat is immediately before exercise, to help prepare an injured or (preferably) recovering muscle for further exertion. Just as a proper warm-up helps to ensure that your muscles and tendons are loose and supple, focused heat at the site of a nagging injury can make sure the affected muscles are as warm as possible before you start using them. A 2005 study in the *Archives of Physical Medicine and Rehabilitation* found that pre-heating calf muscles with a heat pack allowed greater ankle flexion, even without any additional stretching. The heat pack in this case started at about 175°F (80°C) and was applied for 15 minutes;

towels between the pack and the subject's skin ensured that it didn't exceed "comfortably warm." As expected, deep heating using a shortwave machine produced even greater increases in flexibility.

Overall, the clinical evidence on the use of heat suggests a few guidelines: use it before exercise rather than after, and don't use it on a fresh injury. Beyond that, the evidence is thin enough that it comes down to personal preference. It won't hurt you, and—as the Japanese researchers have shown—it might make you feel better.

Will massage help me avoid soreness and recover more quickly from workouts?

There's one big problem with studying the effectiveness of massage: you can't control for the placebo effect. "It's not something where you can do a double-blind experiment," admits Trish Schiedel, past president of the Canadian Sports Massage Therapists Association. So you often see results like the ones published in 2008 by researchers at the Poznan University of Medical Sciences in Poland. They asked volunteers to perform difficult exercises with both arms, then massaged just one of the arms, and evaluated recovery over the next four days. Subjects said the massaged arms felt better—but there were no measurable differences in swelling or range of motion. That's pretty much the story of massage research so far: lots of anecdotal evidence and a scarcity of hard facts. But in the last few years, some studies have emerged that avoid the placebo problem and debunk some old myths.

During the many years that muscle soreness after workouts was thought to result from an accumulation of lactic acid, massage was believed to help flush the acid out. The lactic acid theory is now largely discredited (see p. 60)—and even if it wasn't, experiments published in 2010 by researchers at Queen's University in Kingston, Ontario, show that post-exercise massage actually slows down removal of lactate rather than speeding it up. The reason, according to lead researcher Michael Tschakovsky, is that massage strokes mechanically compress tissue, which squeezes blood vessels shut and prevents the flow of blood.

To find out how massage does work, researchers at Ohio State University enlisted the aid of rabbits, which aren't susceptible to the placebo effect. They first exercised the sedated rabbits by triggering a nerve impulse that causes contractions of a leg muscle. They then used a machine to deliver "cyclic compression forces" that simulated 30 minutes a day of Swedish massage (the most common type of sports massage). The results were clear: massaged muscles regained 59 percent of their lost strength after four days, whereas rested muscles regained only 14 percent. The massaged muscles had fewer damaged fibers and almost none of the white blood cells associated with muscle damage. They also weighed less, suggesting that massage had helped prevent swelling. Interestingly, the results were much less pronounced if the first massage was delayed for a day after exercise, suggesting that the sooner you get your massage, the better.

The rabbit results won't extrapolate perfectly to humans, cautions Thomas Best, the researcher who led the Ohio State study. But these quantifiable outcomes should help scientists begin to figure out the duration, frequency, and strength of the massage stimulus that produces the best results. Right now, the right strength is determined by feel, while frequency and duration tend to be a function of how much you can afford. "If you ask five different therapists, you'll get five different answers," Best says. That means finding a good therapist with expertise in sports massage (not aromatherapy and executive stress backrubs) is essential.

Should I take painkillers for post-workout soreness?

"Non-steroidal anti-inflammatory drugs," more commonly known as NSAIDs, are a class of painkiller whose most famous members are aspirin and ibuprofen. While some NSAIDs are prescription only, others are freely available over the counter, making them the first recourse for many people with exercise-related soreness and pain. A survey of participants at the 2008 Brazil Ironman Triathlon, for example, found that nearly 60 percent had used NSAIDs in the previous three months. Similarly, a quarter of the athletes drug-tested at the Sydney Olympics in 2000 had taken NSAIDs in the previous three days.

The reason they're so popular is that they work—they're highly effective at fighting pain, fever, and inflammation. If you twist your ankle, NSAIDs are the way to go. But that doesn't mean they're the right choice for exercise-related soreness. Some studies have found that NSAIDs do reduce the feeling of post-exercise soreness. What they don't do is speed up the repair of your damaged muscles—and in fact, there's evidence that taking NSAIDs when you're exercising vigorously can slow down recovery and create a whole new set of problems.

Nobel Prize–winning research by chemist John Vane in the 1970s showed that NSAIDs work by inhibiting the production of chemicals called prostaglandins, which regulate inflammation. But prostaglandins also play a role in the creation of collagen, the main building block of musculoskeletal tissue. The soreness and inflammation you feel a day or two after exercise is directly linked to the repairs that make your muscles stronger. In effect, taking NSAIDs to reduce this inflammation cancels out some of your adaptations to training—and indeed, studies have found that prolonged NSAID use delays the healing of bone fractures, as well as injuries to muscles, tendons, and ligaments.

Some athletes also pop NSAID pills before or during vigorous exercise, hoping to ward off later pain. In a study of ultra-marathoners at the Western States Endurance Run, researchers at Appalachian State University found that runners who took ibuprofen before and during the race didn't finish any faster than those who abstained, and they had the same levels of perceived exertion, muscle damage, and soreness in the week following the race. More worryingly, though, blood tests of the ibuprofen users indicated higher levels of endotoxemia, a condition in which toxins leak out of the intestines into the bloodstream. This could be linked to the role of prostaglandins in forming the acid-resistant lining of your stomach and intestine (which is also why NSAID overuse can lead to stomach problems like ulcers and gastrointestinal bleeding).

All these warnings might make NSAIDs seem like dangerous pills that should be avoided at all costs. That's not the case—they're extremely useful when used in the right context. That means they're not suitable for dealing with nagging, chronic problems on an ongoing basis, or for trying to prevent pain before it happens. But if you have an injury that swells and is painful even when you're not exercising, NSAIDs, for no more than seven days, may be just what your doctor (or therapist) orders.

How long does it take to recover after a marathon or other long, intense effort?

The aches and pains that hit you in the days after an extended bout of exercise can seem worse than the competition itself. When you take part in something like a triathlon, a multi-hour hike, or a long running race, you're subjecting your body to stress that causes damage and takes time to heal. Different systems return to normal at different rates: acute fatigue might be gone within a day or two; your immune system could be weakened for up to a week; and in some cases, muscular fatigue can linger for several weeks.

In the last few years, researchers have become concerned about the possibility that extended endurance activity could cause heart damage, since the heart has to beat unusually quickly for several hours. Several studies have found evidence of "cardiac injury" in runners after they complete a marathon, including enzymes suggesting that heart muscle has been damaged. To investigate these claims more closely, researchers at the University of Manitoba used magnetic resonance imaging to perform a detailed analysis of the hearts of participants in the Manitoba Marathon, publishing the results in the *American Journal of Cardiology* in 2009. They found that, despite initial evidence of damage, normal heart function resumed within a week. In other words, your heart muscles take a beating during a marathon but recover soon afterwards, just like your leg muscles.

The damage to your legs will be considerably more obvious to you, since you may have difficulty walking down stairs or even just getting out of bed the next morning. This delayed onset muscle soreness, or DOMS (see p. 61) tends to peak one or two days after the race and can persist for up to a week. Soreness lasting longer than a week could indicate more serious damage and should be examined by a clinician. But it's not unusual for muscle fatigue to persist even after the soreness is gone, sometimes for several more weeks. A Danish study in 2007 tested well-trained runners a week after a marathon, when soreness was no longer a factor. Using electrodes to stimulate muscle contractions, the researchers found that the muscles themselves had fully recovered—but when the runners tried to voluntarily contract their muscles, they were still much weaker than before the race. This suggests that the lasting fatigue after a marathon has a neuromuscular origin—the signal from the brain to the muscle fibers is disrupted somewhere along the signal path (one theory suggests the disruption is caused by overloaded receptors in the spine).

Despite their best efforts, researchers haven't had much success in figuring out how to speed up the recovery process. A classic study in 1984 compared experienced marathoners who took a week of complete rest to those who ran 20 to 45 minutes a day after the race. After a week, the rested group had better leg muscle strength and slightly higher levels of energy storage in their muscles, though the differences weren't large. Other studies with similar results suggest that it's best to be cautious and make a gradual return to activity. Start with walking (or gentle biking or swimming) instead of running during the first four or five days. After that, proceed with a "reverse taper" that reaches normal training no sooner than two weeks after the race. And be flexible: if your legs still feel dead after three weeks, congratulate yourself on having pushed very close to your limits in the race—and give them more time to recover.

Can "platelet-rich plasma" cure my tennis elbow or Achilles tendon?

One of the minor subplots in the media frenzy that engulfed Tiger Woods in late 2009 was his connection to a Toronto sports doctor named Anthony Galea, who said he had injected Woods at least four times with "platelet-rich plasma," or PRP, to help his recovery from knee surgery. Further investigation revealed a long list of prominent athletes from Olympic sports and almost every major professional league who had received PRP therapy during injury rehab.

For the general public, the investigation offered a rare peek into the world of cutting-edge, sometimes experimental sports medicine treatments that top athletes rely on. It also offered hope for recreational athletes sidelined by chronic tendon injuries—was it possible that a simple, non-invasive procedure might heal them? In fact, Galea reported that about 40 percent of patients seeking the treatment from his clinic were recreational rather than professional athletes. But researchers are still hotly debating whether the technique actually works, and if so, how it should be used. (Despite rumors to the contrary, PRP is not illegal; however, since the beginning of 2010, elite athletes who are subject to drug testing have had to apply for a "therapeutic use exemption" before receiving PRP therapy, and injections of PRP directly into muscle tissue are banned.)

The technique is designed to help injuries that don't heal well on their own. For example, unlike muscles, tendons have a very poor blood supply, so minor tears and inflammation tend to heal slowly. PRP therapy involves drawing a small amount of the patient's own blood, spinning it in a centrifuge to concentrate the most useful components (the platelets), and then re-injecting this concentrated plasma at the injury site. The platelets then release various "growth factors" that stimulate the body's natural healing response.

The technique isn't new—the first attempts to harness the growth factors in platelets date back at least to the early 1980s, and surgeons have also experimented with PRP to aid in bone grafts. But it's only in the last five years or so that small pilot studies have shown the technique's potential for tendon injuries. Initial results for Achilles tendinopathy and tennis elbow—two stubborn tendon problems that often resist non-surgical treatment—were promising, but the studies were neither randomized nor placebo-controlled.

Only now are the first proper clinical trials emerging, and the verdict remains unclear. A team of researchers in the Netherlands tested the technique on patients with chronic Achilles tendon problems, publishing the results in 2010 in the *Journal of the American Medical Association*. Fifty-four patients received an injection of either platelet-rich plasma or a saline placebo, then undertook a program of rehab exercises. After 24 weeks, the two groups were indistinguishable, dealing a blow to hopes that PRP would prove to be a "magic bullet."

The next month, a different Dutch team published another clinical trial, this time with positive results. In a group of 100 patients with tennis elbow, 73 percent of those given PRP reduced their pain by at least 25 percent after a year, whereas only 49 percent of those who received a corticosteroid injection achieved similar results. Again, the study was blinded so that the patients didn't know which procedure they were receiving.

The shots typically cost about $500 each, and a full course of treatment including imaging can exceed $2,000. With that in mind, you can balance the benefits of a promising but unproven treatment against the cost—and how long you've been struggling with the injury. It's worth noting that, even in the unsuccessful trial, both the PRP and placebo groups did improve significantly during the study (by about 20 points on a 100-point pain scale). The authors note that "placebo response is amplified when a treatment is invasive and raises high expectation." In other words, the shots could work if you believe in them.

How can I reduce my risk of stress fractures?

Evan Lysacek had one in his left foot the year before winning the 2010 Olympic gold in men's figure skating. Tiger Woods had two in his left leg. A *Globe and Mail* reporter even got one from dashing between parties at the Toronto International Film Festival in high heels. Stress fractures are among the most common—and dreaded—diagnoses in athletes. Striking most often in sports that involve running and jumping, they usually signal the end of an athlete's season, since the only treatment is rest for 8 to 10 weeks or more.

Bone is a living tissue, in a constant balance between breakdown and repair. When the damage from repeated impacts builds up more quickly than it can be repaired, microscopic cracks begin to form. Eventually, these cracks join together to form a stress fracture—a hairline crack in the bone that results from weeks or months of accumulated pounding rather than from a single traumatic blow. The single most important factor in preventing stress fractures is having strong, healthy bones. But new research suggests a couple of other factors that might reduce your risk.

The first comes from researchers at Iowa State University, who used a computer model of bone damage and repair to estimate the effects of changing your stride length. Basically, if you shorten your stride, you'll have more foot-strikes per mile (and thus more impacts jarring your bones), but each foot-strike will be a little gentler. So which effect predominates? The researchers had 10 runners run on a treadmill with different stride lengths, measuring the relevant forces with motion-capture cameras and force plates, then plugged the data into their computer model. The conclusion: shortening your stride length by 10 percent reduces stress fracture risk by 3 to 6 percent.

Changing your running stride definitely isn't easy (see p. 88). But other studies have found that one of the chief sins of inexperienced runners is overstriding: elite runners tend to take about 180 steps a minute regardless of how fast they're running, while less experienced runners take fewer steps. So focusing on taking short, quick steps could have multiple benefits—including lowering your stress fracture risk.

The second factor comes from a study of 39 female runners, half of whom had a history of stress fractures. Researchers at the University of Minnesota took a series of measurements to determine the size, structure, and density of the subjects' bones and muscles. Not surprisingly, the shin bones in the stress fracture group were smaller by 7 to 8 percent, and weaker by 9 to 10 percent. Interestingly, though, the bone differences were exactly in proportion to the size of the calf muscles, and there was no difference in bone mineral density.

This suggests that the women in this group who suffered from stress fractures weren't guilty of not getting enough calcium—instead, the relative weakness of their bones was a response to the lack of muscle in their legs. And the fix is simple: strengthen your calf muscles by doing exercises like calf raises. The extra muscle should help cushion some of the impact when you run and jump and also stimulate your shin bones to get stronger. (And this doesn't apply only to your shins: the

best way to ensure the bones throughout your body are strong
is to keep the surrounding muscles strong.)

Should I exercise when I'm sick?

The answer to this question seems obvious: if you're sick, your
body needs its strength to fight off the infection. But exercise
is a deeply entrenched habit for many people, so when illness
strikes, they want to know if they can exercise without doing
themselves harm.

For any sort of serious illness, there's no doubt that you
shouldn't exercise. The question usually arises with less serious
conditions like colds, which are unpleasant but not debilitating.
Although there isn't a great deal of research on the topic, many
researchers apply a rule of thumb known as the "neck check,"
according to Thomas Weidner, the head of the athletic training
program at Ball State University in Indiana. Patients are generally free to exercise if their symptoms are above the neck, like
a runny nose, sneezing, or a scratchy throat. But symptoms
below the neck like fever, aching muscles, or a chest cough are
grounds for caution.

Weidner was responsible for a couple of unusual studies
in the late 1990s in which volunteers were infected with rhino-
virus, better known as the common cold, in one of the very few
attempts to address this question in a controlled experiment.
First, he infected 45 volunteers, who began to develop sore
throats the next evening and proceeded to full-blown symptoms by the third day of the experiment. At the peak of their
illness, he put them through a series of treadmill tests and compared the results to a group of uninfected controls. To his surprise, he found no difference between the two groups in their
running performance, lung function, or any other physiological

responses. In other words, having a cold doesn't seem to make you a worse athlete.

In the second study, Weidner infected 50 volunteers and had half of them do 40 minutes of exercise at 70 percent of maximum heart rate every second day, while the other half just rested. There was no difference between the two groups in the severity and duration of the symptoms—and in fact the exercise group reported feeling slightly better than the controls. Though they're more than a decade old, those results haven't been contradicted by any studies since, Weidner says. (No doubt it's challenging to assemble a group of volunteers willing to be infected with a cold!)

There's plenty of anecdotal evidence to support Weidner's finding that light exercise when you have a cold makes you feel better. People believe it clears the airways, or that the enhanced circulation speeds healing, or that it simply feels good. It's well established that moderate exercise boosts immune function (see p. 23)—and one study even found that a single 45-minute treadmill run helped mice battle a virus. So it's not that far-fetched to believe that staying active while sick might have real physical benefits. For now, though, we'll have to be content with Weidner's finding—that at the very least, exercising with a cold doesn't make your symptoms worse.

Will having a few drinks affect my workout the next day?

That depends on what you mean by "a few."

In 2010, researchers in New Zealand published a surprising study that found significant delays in muscle recovery when the subjects drank a "moderate" amount of alcohol after a strenuous workout. The subjects did a series of leg exercises, then had 90 minutes to drink either straight orange juice or a mix of vodka and orange juice before going to bed. Over the next three days, the alcohol group didn't report feeling any additional leg soreness compared to the oj group—but their loss of strength in a series of tests was 1.4 to 2.8 times greater.

However, "moderate" in this case was 1 gram of ethanol per kilogram of body weight, corresponding to about 6.5 bottles of 5-percent-alcohol beer for the subjects, who had an average weight of 193 pounds (87.6 kilograms). "When you look at how much athletes are reported to drink in the scientific literature, this is actually a moderate dose," says Matthew Barnes, a researcher at Massey University and the lead author of the study. "This is not just in New Zealand but in the majority of Westernized countries where contact team sports are played."

Sure enough, researchers have recorded some fairly prodigious feats of drinking by athletes—Barnes points to one study of rugby players in which post-match consumption reached as high as 38 units of alcohol, or 22 bottles of beer. Since most people are dealing with smaller quantities, Barnes ran a follow-up study that cut the dose in half, to 0.5 grams of alcohol per kilogram of body weight. The results, published in the *European Journal of Applied Physiology*, offer good news: he found no difference at all in recovery between the alcohol and orange juice groups.

He and his colleagues are now conducting further experiments that suggest the higher dose of alcohol may affect the central nervous system rather than the muscles themselves, weakening the signals sent from the brain to the muscles. However, it's possible that there are also changes in the muscles, or in the levels of hormones like cortisol and testosterone.

Two other key factors affect how well you recover after a workout-booze combo, Barnes says: rehydration, and refilling your carbohydrate stores. Drinks containing more than about 4 percent alcohol have a diuretic effect; drinking a standard shot of hard alcohol will make you expel four times that much urine. The solution here is simple: drink a glass of water for every alcoholic drink you have during the evening.

In order to recover properly after a workout, replenishing your energy stores during the two hours following exercise is crucial. Some animal studies have suggested that alcohol can directly hinder your ability to restore carbohydrate levels, but these results remain disputed. However, problems definitely arise if the calories you consume from alcohol displace more functional calories. A 2003 study of Australian cyclists found that simply adding alcohol to a post-workout meal didn't change the amount of carbohydrate stored. But if the alcohol replaced some of the calories in the post-workout meal, carbohydrate stores were 50 percent lower after eight hours, and still lower 24 hours later.

Overall, these studies fit with the prevailing wisdom that one or two drinks a night won't have any negative effects on your health and performance. Indeed, light to moderate drinkers appear to have 20 to 40 percent lower risk of heart disease, among other reported benefits. But if you've arranged a big night out with a group of Kiwi rugby players, you might want to schedule a fairly light workout for that day—or at least, don't expect to set any personal bests in the days that follow.

CHEAT SHEET: INJURIES AND RECOVERY

- "RICE" (rest, ice, compression, elevation) is important immediately after soft-tissue injury, but after acute swelling has passed switch to "MICE" (mobilization, ice, compression, elevation) to avoid scar tissue build-up.
- Ice baths may help speed recovery from muscle soreness, using bouts lasting at least five minutes and temperature of 50°F (10°C).
- Heat packs can loosen tight or injured muscles, but only if they're near the surface. Use heat before exercise to aid warm-up, not after.
- Massage doesn't flush away lactic acid but may speed recovery from muscle soreness. Use a practitioner who specializes in sports massage.
- Anti-inflammatory drugs like aspirin and ibuprofen are not suitable for chronic, nagging injuries or to prevent pain before it happens. They carry health risks and may interfere with the effects of training. However, they're suitable for acute injuries.
- After an extreme event like a marathon, your body will return to normal within about a week, but neuromuscular fatigue can persist for several weeks.
- "Platelet-rich plasma" is a component of your own blood, injected to speed healing of tendon injuries. Recent clinical trials suggest it's not a "miracle cure," but it may speed healing in some patients.
- The best way to keep your bones strong enough to avoid stress fractures is to strengthen the muscles around them. Shortening your running stride may also help.
- For "above-the-neck" symptoms like a runny nose or a sore throat, exercising with a cold appears to have no ill effects, and may even speed recovery slightly.
- Having a few drinks won't affect your next day's workout, but more than four or five (depending on your weight) can slow muscle recovery and displace other needed nutrients.

8: Exercise and Aging

ATTITUDES TOWARD EXERCISE AND AGING have changed dramatically in recent decades, as shown by 73-year-old Ed Whitlock's record-setting sub-three-hour marathon in 2004. His time of 2:54:48 wasn't just fast for an old guy—he placed 26th out of more than 1,400 finishers. The aging body is capable of much more than we once believed, but study after study has shown that we have to "use it or lose it." As a result, researchers are busy figuring out what kinds of exercise are best for keeping our bodies and minds young.

For athletes like Whitlock, though, the biggest question is not what exercise can do for their aging bodies, but what aging will do to their 5K times. "Masters" sport (often defined as over-40) is the fastest-growing segment of sport in North America, and the experiences of these remarkable athletes offer valuable lessons about how to stay motivated and adjust workout routines as we age.

What's the cumulative effect of all the exercise I've done over the years?

Unless you make a dramatic turnaround after a severely misspent youth, it's inevitable that some of your body parts won't work as well in your 50s as they did a few decades earlier. It may be tempting to blame that on the punishment you've inflicted on your body during years on playing fields, ice rinks, and jogging paths—but the opposite is more likely. In fact, researchers have a good idea of the average rates of decline you can expect for various systems (see p. 166). And for almost every sign of aging you can think of—muscle loss, weight gain, artery hardening, joint stiffening—there have been studies suggesting that exercise slows it down.

It's not just the obvious physical ailments that exercise fights off. Better circulation of blood to the brain helps delay mental decline, and at a microscopic level exercise appears to slow the aging of your cells. Some benefits aren't yet fully understood, like the 2009 finding by Lawrence Berkeley National Labs researcher Paul Williams that aerobic exercise dramatically reduces the risk of glaucoma, macular degeneration, and cataracts. This may have something to do with links between cardiovascular fitness and fluid pressure behind the eye—but in a sense, it doesn't really matter how it works. The important thing is that, thanks to epidemiological studies, we know that exercise is the most powerful anti-aging tactic we've got.

Of course, many competitive sports do result in some wear and tear, and a series of studies have connected sports like soccer and hockey to elevated risk of knee osteoarthritis in later life. More recent studies, though, have differentiated between the risk of simply playing these sports and the risk that arises from acute knee injuries. According to a Swedish study in 2006, the increased risk of arthritis for soccer and hockey players was entirely attributable to those who had suffered serious knee injuries during their playing career. Moreover, a 2008 study of former top-level Tunisian soccer players aged 45 and over who had never suffered acute knee injuries found that they had less knee pain and less functional disability than a group of matched controls.

THE AGING BODY

Regular physical activity has tremendous power to slow the effects of time. Still, even the American College of Sports Medicine, in its statement on exercise and physical activity, concedes that "no amount of physical activity can stop the biological aging process." These are some of the key changes you'll notice.

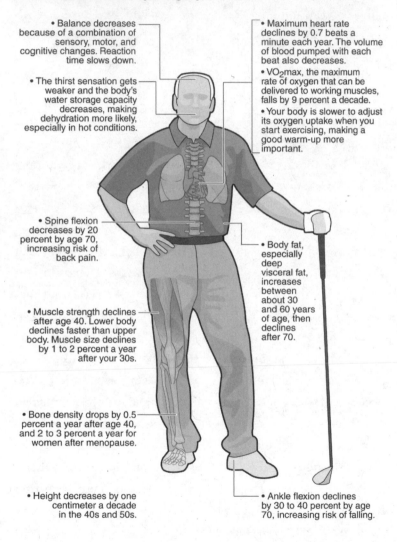

• Balance decreases because of a combination of sensory, motor, and cognitive changes. Reaction time slows down.

• The thirst sensation gets weaker and the body's water storage capacity decreases, making dehydration more likely, especially in hot conditions.

• Maximum heart rate declines by 0.7 beats a minute each year. The volume of blood pumped with each beat also decreases.

• VO$_2$max, the maximum rate of oxygen that can be delivered to working muscles, falls by 9 percent a decade.

• Your body is slower to adjust its oxygen uptake when you start exercising, making a good warm-up more important.

• Spine flexion decreases by 20 percent by age 70, increasing risk of back pain.

• Body fat, especially deep visceral fat, increases between about 30 and 60 years of age, then declines after 70.

• Muscle strength declines after age 40. Lower body declines faster than upper body. Muscle size declines by 1 to 2 percent a year after your 30s.

• Bone density drops by 0.5 percent a year after age 40, and 2 to 3 percent a year for women after menopause.

• Height decreases by one centimeter a decade in the 40s and 50s.

• Ankle flexion declines by 30 to 40 percent by age 70, increasing risk of falling.

If it's acute injuries rather than wear and tear that lead to arthritis, you might expect running to be in the clear—and indeed that's what a series of recent studies have concluded. Needless to say, this conclusion will be hard for many people to believe. After all, the aging runners they know are certainly subject to aches and pains. But the data collected by Williams and others suggest that, while everyone acquires aches as they age, it's the people who don't exercise who acquire the most. "As these runners aged," Williams noted after one of his studies, "the benefits of exercise were not in the changes they saw in their bodies, but how they didn't change like the people around them."

Will running ruin my knees?

This is a fear that stops many would-be runners in their tracks and lurks in the back of the mind of even the most experienced. Occasional aches and pains are pretty much an inevitable part of running on a regular basis, so it's entirely reasonable to wonder whether the exercise you're enjoying now will leave you hobbling in a decade or two. Over the past few years, several long-term studies have produced results that should put these fears to rest.

In a 2008 issue of *Skeletal Radiology*, a team of Austrian radiologists presented knee MRIs of seven runners who had taken part in a previous MRI study before running the Vienna marathon in 1997. The use of MRIs offers a significant diagnostic advantage compared to earlier studies that relied on X-rays. The results were clear: no new damage in the knee joints of the six subjects who had continued running in the intervening decade. "In contrast, the only person who had given up long-distance running showed severe deterioration in the intra-articular structures of his knee," the authors note.

An even longer-term study at Stanford University has been following 45 runners and 53 non-running controls since 1984, taking regular X-rays. The latest results, which appeared in the *American Journal of Preventive Medicine* in 2008, showed that after 18 years, 20 percent of the runners had developed osteoarthritis (the most common form of arthritis) in the knee, compared to 32 percent of non-runners.

These two studies raise a possibility that several earlier studies have proposed: that, far from ruining your knees, running might actually help protect them. Due to the limited data available, it's not possible to draw that conclusion at this point, Stanford lead author Eliza Chakravarty cautions. "I don't think I would strongly recommend running for the purpose of 'protecting the knees,'" she says. Still, the idea is plausible: the American College of Sports Medicine recently reported that each additional pound of body mass puts four additional pounds of stress on the knee, so packing on a pound a year for a decade ups your chances of developing arthritis by 50 percent—a fairly powerful argument for running to protect your knees.

One important drawback with both studies is selection bias. The runners in both studies were committed recreational runners who already had a history of being able to run without serious problems. A more rigorous test would involve testing a random sampling of the general population, rather than pitting "runners" versus "non-runners." That's effectively what researchers from the famously long-running Framingham Heart Study did, analyzing data from 1,279 subjects over a nine-year period and publishing the results in the journal *Arthritis & Research* in 2007. Using the comprehensive medical and lifestyle data accumulated for the study, the researchers found no association between exercise (including running) and the development of knee osteoarthritis.

Of course, the decision doesn't have to be strictly utilitarian. As one of the Vienna study participants (who was preparing to run his 37th marathon) put it in an e-mail to lead author Wolfgang Krampla, "Even if minor aches and pains occur over the years, the gain in 'joie de vivre' far outweighs them."

How should I adapt my workout routine as I get older?

One of the recurring themes in coverage of the 2008 Olympics was that old people can be just as strong and fast as their juniors. At 41 years of age, swimmer Dara Torres won three silver medals; marathon runner Constantina Tomescu-Dita won gold at 38; and 61-year-old Ian Millar picked up a remarkable silver medal in the team equestrian event. But it's not entirely clear what lessons the average middle-aged or elderly exerciser can draw from these one-of-a-kind models.

Probing that question by studying "masters" athletes—the definition varies from sport to sport, but it often refers to ages 40 and over—has become a hot research topic in recent years, in part because masters competition is the fastest-growing segment of sport in North America. "We're not trying to encourage everybody to become a masters athlete," says Patricia Weir, a professor of kinesiology at the University of Windsor. "Most adults simply won't choose to undergo that level of training." Instead, Weir and collaborators like Bradley Young of the University of Ottawa are trying to figure out what key factors allow some athletes to train and compete at a high level for many decades, to see whether there are insights that could help weekend warriors stay active as they age.

The basic principles of training for older athletes are the same as for younger athletes, according to a review of the topic by University of Wisconsin-La Crosse sports scientist Carl Foster and his colleagues in 2007. However, the optimal mix of stimulus and recovery may be shifted by the risk of injury, as well as by the hormonal changes that accompany aging, which affect the speed and magnitude of the body's response to exercise. To combat the steady loss of muscle with age, for example, Foster recommends weight training, even—or perhaps especially—for skinny endurance athletes.

Studies of masters athletes have found that the most successful manage to suffer fewer injuries than their peers. This may seem like a matter of luck, but it's not necessarily that simple. "Maybe they've got good genetics," Young says, "but maybe they're also smart." You can improve your odds through what Young calls "deliberate acts of recovery," such as taking an extra day between hard workouts. Cross-training may also be more valuable for older athletes, since they're less able than younger athletes to recover from doing the same activity every day.

Champion masters athletes continue to train intensely, Young and others have found, but their training becomes more focused on the essentials. Top age-group distance runners, for example, spend proportionately more time training their endurance as they age. But the dominant factor appears to be consistency and continuity of training. Top masters athletes manage to accumulate months, years, and even decades of relatively unbroken training. Although champions train for their chosen sport year round, many recreational masters athletes prefer "sampling": they participate in different sports throughout the year but are never totally inactive for long stretches. Ultimately, this may be the most important lesson we can draw from the exploits of aging Olympians: if you play hockey for six months a year, find something else to keep you active for the other six months.

How quickly will my performance decline as I age?

One way to see how human performance declines over time is to plot the age-group world records for various sports. No matter what sport you choose, you'll see a similar pattern: gradual declines starting around the age of 35, getting steadily steeper with each succeeding decade. For a typical healthy adult, each decade past your 30s brings on average a 9 percent decrease in aerobic fitness (measured by the maximum amount of oxygen you're able to process), a drop of seven beats in your maximum heart rate, and the loss of 10 percent of the muscle in your body.

You don't lose everything at the same rate. If you compare the age-group world records for the 100 meters and the marathon, you see a sharper decline in endurance compared to speed. This pattern is replicated in other sports like swimming. The decreases, as a percentage of the world record, also appear to be steeper in women than men—but researchers suspect that there's no physiological reason for this. Instead, the records for older women are likely weaker because not as many women as men continue to compete in organized sports as they get older.

Another study, by researchers at the University of Texas's Cardiovascular Aging Research Laboratory, examined age-group records for weightlifting. The records for Olympic-style lifts like the snatch and the clean-and-jerk, which rely more on muscular power than on absolute strength, declined more quickly than pure tests of strength like the bench press, squat, and deadlift. It's this loss of muscular power, rather than strength, that causes the most problems for seniors in day-to-day life—which is why researchers now recommend that seniors include at least some power-building exercises in their program (see p. 108).

Age-group records offer a general picture of how the human body changes over time. But if you're interested in knowing how your own body will change, these records have a subtle but important flaw. The records show the best performances of many different people, each of whom flourished for a brief period before fading to sub–world-record levels at later ages, thanks to health problems, changes in motivation, or other issues. There's a big difference in the performance trends displayed by this type of "cross-sectional" data compared with "longitudinal" data that follow specific individuals for many years—and the differences are encouraging.

Starting with a study of masters track and field athletes published in *Experimental Aging Research* back in 1982, researchers have found that longitudinal data show a less steep decline in performance than cross-sectional data (i.e., age-group records). The decline seen in age-group records tends to be quadratic, which means it gets steeper and steeper as years pass. On the other hand, more recent studies of runners and swimmers who have trained continuously for several decades show evidence of decline in a straight line. You'll get slower no matter what, but continuous training seems to prevent the decline from accelerating (at least for a while).

For sports like running and swimming, age-graded performance tables offer a way to assess how you're doing from year to year. For each five-year age category, performances are expressed as a percentage based on statistical analysis of existing performances. Just remember that if you can stay healthy, motivated, and consistent, you should be able to beat this cross-sectional curve—if you do, your age-graded score will get better as you get older.

SLOWING DOWN

Comparing the age-group records over 100 meters and the marathon suggests that endurance declines more steeply than speed.
Women also decline more steeply than men, though that may result from lower participation levels in older age groups.

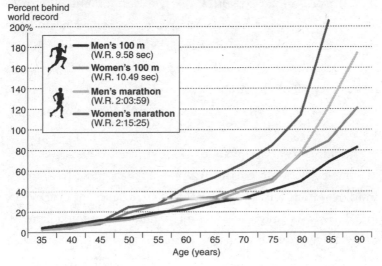

Percent behind
world record

Legend:
- Men's 100 m (W.R. 9.58 sec)
- Women's 100 m (W.R. 10.49 sec)
- Men's marathon (W.R. 2:03:59)
- Women's marathon (W.R. 2:15:25)

Age (years)

How can I stay motivated to exercise as my performances decline?

When Ed Whitlock became the first septuagenarian to run a marathon in under three hours in 2004, it was thanks to a simple but grueling training plan consisting of two-to-three-hour runs around a local cemetery, nearly every day. That regimen presented two key challenges that are familiar to any masters athlete: staying healthy, and—just as important but less obvious—staying motivated. In fact, when asked why his race performances in his 50s were less impressive than the years before and after, Whitlock points to the second factor rather than the first. "The main reason—or excuse—is that I was busy at work, so while I did continue running throughout the decade, my training dedication fell off," he says. "I am sure if I had been more dedicated and better organized I could have done more."

The physical declines that accompany aging are well documented (see p. 166), but there's strong evidence that the decline in masters athletic performance is steeper than purely physiological reasons can explain. Indeed, experiments where mice are given unfettered life-long access to exercise wheels have suggested that the intrinsic drive to exercise declines with age. In a sense, it's a classic chicken-and-egg problem: Do you train less as you get older because you're slower and weaker, or is it the other way around? The answer seems to lie somewhere in the middle.

In this context, motivation seems crucial to success. Studies of elite masters athletes by University of Ottawa researcher Bradley Young and his colleagues have identified a complex mix of personal and social factors that make some people more likely to continue training at a high level into their 50s and beyond. The personal factors are mostly what you'd expect: at the top of the list is enjoyment of the sport, which is cited by more than half of elite older athletes as a reason to train. Close behind is the sense of personal challenge, followed by improved fitness and health. Surprisingly, the social factors mix positive reinforcement (from family, training partners, and the wider community) with more negative pressure (the feeling, for instance, that stopping training would make you a quitter).

Motivation is ultimately very personal, so there's no universal formula for maintaining your enthusiasm. But it's important to understand how the people around you can affect your outlook and to make sure that your family and friends are supportive and understand the benefits of your exercise routine. As researchers studying three decades of data from the Framingham Heart Study have found, health and exercise habits are highly contagious. And the most powerful source of social pressure, Young's research shows, is your spouse—a result that wouldn't surprise Whitlock, who discovered his calling as a masters runner when he was 40, after a 15-year hiatus, at the urging of his wife.

What are the pros and cons of exercising in water?

One of the classic problems with exercise is that those who most need its health benefits sometimes have the greatest difficulty doing it. If you have osteoarthritis, your joints hurt; if you're battling obesity, the impact forces will put you at risk of injury; if you're a senior, you could break bones in a fall. A common solution to this problem is to exercise in the water, often in the form of water aerobics or aquafit group classes. There's no doubt that this lower-impact approach helps reduce some of the risks of exercise for at-risk groups, but it's only recently that researchers have started to ask whether it really provides the same benefits as land-based workouts.

One key difference between water aerobics and swimming stems from the fact that you're standing vertically in the water, rather than floating horizontally. Since water pressure underwater is greater than it is near the surface, the pressure exerted on your feet will be greater than it is on your chest. As a result, blood is pushed back from your extremities toward your heart with less effort than usual. This means that your heart rate will be lower at a given level of effort—meaning you have to work harder to get the equivalent cardiovascular workout that you'd get on land.

Nonetheless, a number of studies have found that water-based exercise has valuable benefits, particularly for groups with special needs. For example, a 2009 study found that a four-week aquatic exercise program consisting of a mix of aerobic, stretching, and strengthening exercise produced greater improvements in patients with lower-back pain than a comparable four-week program on land. The researchers speculated that water pressure and temperature reduced pain signals during exercise, and buoyancy reduced stress on joints and muscles, enabling a greater range of motion.

Several studies have explored aquatic exercise for patients with hip and knee osteoarthritis. A review of these studies in 2009 concluded that it likely produced short-term benefits, though further research is required to determine whether the benefits persist in the long term. More generally, older adults often face a combination of different challenges—aching joints, poor balance, weak bones—that make water a more attractive option. A 2008 study of 50 women between 62 and 65 years old found that water-based exercise produced greater improvements in cardiovascular fitness, agility, and flexibility than a land-based walking program. Of course, it's not really fair to compare a water-based program that includes aerobic, strength, and flexibility exercise to a simple walking program. The key difference isn't that aquatic exercise is better than exercise on dry land—it's that people feel comfortable doing exercises in the pool that they wouldn't do at all on land.

There is, however, one recent study that hints at the possibility that training in water might offer some subtle benefits that you can't duplicate elsewhere. In 2009, French researchers found that a group of patients with either chronic heart failure or coronary artery disease displayed an unexpected rise in nitric oxide levels in their blood vessel linings after a three-week aquatic exercise program—a change expected to lower their risk of death. Although the findings are very preliminary, the researchers suggest that the altered circulation caused by the water's pressure gradient may produce additional cardio-vascular effects (beyond the lowered heart rate noted above). For now, though, it's just a theory.

What type of exercise is best for maintaining strong bones?

The key word here is "maintain," since 95 percent of your mature skeleton is already in place by the age of 17 for girls and 19 for boys. Once you reach adulthood, it's basically one long fight against the slow but inexorable weakening of your bones. According to conventional wisdom, the key to that fight is en-gaging in weight-bearing activities—those in which you're stand-ing and supporting your own weight rather than being seated. But the latest research shows strength training can also play a key role—and in fact, lifting weights may be even more effective than some weight-bearing activities like elliptical training.

"Over the past decade, people have realized that bone is more dynamic than we thought. It's actually a pretty respon-sive tissue," says Heather McKay, a professor in the faculty of medicine at the University of British Columbia and the director of the Centre for Hip Health and Mobility. It turns out that train-ing your bones has more in common with training your muscles than previously thought: if you stress them, they'll get stronger. How much stronger depends on what your body is currently used to, how big a load you apply, and how you apply it. Recent studies by McKay's team have found that short bursts of intense activity separated by brief rest periods—anything from jump-ing on the spot to squats in the weight room—build bone more effectively than continuous, less intense activities.

This means that weight bearing, on its own, is a bit over-rated. It's true that the skeleton gets a bit of a workout from gravity whenever you're standing up, but you can stress your bones in a more targeted manner by training with weights. "Any time you're increasing your muscle mass, the tension of the muscles on the bone creates a 'bending moment' that stimulates your bones," McKay explains. Lifting weights also allows you to target vulnerable areas like your wrists, which get no benefit from your hours on the elliptical.

Another study by McKay's group found that schoolchildren who jumped up and down between 5 and 15 times, three times a day (at the morning, noon, and end-of-school bells) significantly increased their bone density. Since a quarter of your adult skeleton is laid down during early puberty, it's important to make sure children are doing the kinds of activities that build strong bones—and this study confirms that even small amounts of intense, jarring activities like jumping are more effective than simply standing or walking around.

Numerous studies over the years have found that strength-trained athletes have greater bone mineral density than endurance-trained athletes, lending support to the idea that building muscle is better for bones than weight-bearing activities like running. But a 2009 article in the *Journal of Strength and Conditioning Research* showed that the differences aren't that simple. Pamela Hinton and her colleagues at the University of Missouri compared runners, cyclists, and strength-trained men. They did find that the strength group had the greatest bone density, but that was only because they had the biggest bodies. The runners were leaner, but their bones were just as strong relative to their body size.

There was, however, a significant difference between the bone density of runners and cyclists, which suggests that it's the repeated, jarring impacts of running that produce stronger bones compared to cycling. As a result, Hinton recommends

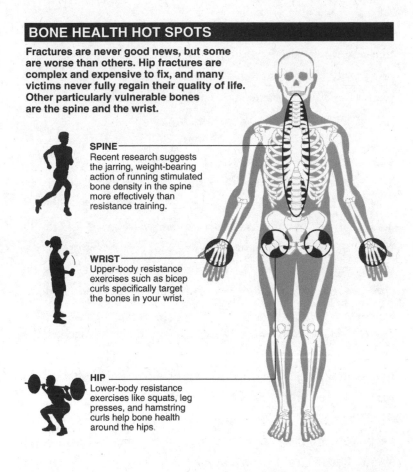

BONE HEALTH HOT SPOTS

Fractures are never good news, but some are worse than others. Hip fractures are complex and expensive to fix, and many victims never fully regain their quality of life. Other particularly vulnerable bones are the spine and the wrist.

SPINE
Recent research suggests the jarring, weight-bearing action of running stimulated bone density in the spine more effectively than resistance training.

WRIST
Upper-body resistance exercises such as bicep curls specifically target the bones in your wrist.

HIP
Lower-body resistance exercises like squats, leg presses, and hamstring curls help bone health around the hips.

that those who engage in activities such as cycling, swimming, and rowing consider adding a dose of either strength training or a higher-impact activity like running to their regimen. That also means that elliptical trainers, which many people turn to precisely for their softer landing, suffer from the same shortcoming. "There's no impact force, as the steps of the machine move with you," Hinton says.

Sports like soccer and basketball (and activities like step aerobics) offer the best of both worlds, stimulating bone health through the impact of intermittent jumping and running, as well as by building muscle strength. Hinton's research suggests that you don't necessarily have to lift weights, and you don't necessarily have to run and jump either—but you need to do one or the other to make sure you're either building muscle or getting jarring impacts.

Can exercise keep my DNA from aging?

The 2009 Nobel Prize in medicine was awarded to three researchers who discovered how DNA can be copied over and over again without being damaged. Telomeres are short lengths of DNA at the end of each chromosome that serve as a protective cap, ensuring that the delicate ends aren't snipped when the chromosome is replicated. Unfortunately, the telomeres themselves get shorter and shorter as you get older—and once they reach a certain minimum, the cell has effectively reached the end of its life. In fact, some researchers now believe that telomere shortening is the fundamental change that underlies all aging: as your cells age, so do you.

Exercise, we've been told for years, is a fountain of youth. It keeps your arteries supple, your muscles strong, and your mind sharp. But it hasn't been entirely clear how it accomplishes such diverse effects. A 2010 paper by University of Colorado researchers, published in the journal *Mechanisms of Ageing and Development*, offers a clue. Physiologists analyzed the telomere length and aerobic fitness of four groups of people: young (18 to 32) and sedentary, young and fit, old (55 to 72) and sedentary, old and fit. The "fit" subjects did at least 45 minutes of vigorous exercise five times a week. The two young groups had essentially the same length of telomeres. The telomeres of the old, fit group were slightly but not significantly shorter. But the old, sedentary group had dramatically shorter telomeres.

The Colorado researchers also plotted telomere length versus aerobic fitness (VO_2max) for the older group. Across the board, those with higher fitness had longer telomeres. The implication is clear: vigorous aerobic exercise makes your DNA look several decades younger than it is. And that's bad news for the sedentary group. A recent study of 780 heart disease patients, for example, found that those with the shortest telomeres were most likely to die in the following four years—a mortality risk that couldn't be explained by other known risk factors in these patients.

It's important to note that the Colorado study can't distinguish correlation from causation. It's possible that a hidden underlying factor makes some people keep longer telomeres and also drives them to exercise. But another study by German researchers from the University of Saarland suggests that this isn't likely. Like the Colorado study, the Saarland study found that older runners and triathletes had telomeres almost as long as younger subjects, while older non-exercisers had much shorter telomeres. In addition, the German study tested mice who were assigned to either run on an exercise wheel or do no exercise. After just three weeks, the exercising mice displayed higher levels of telomerase, the enzyme that stimulates the formation of telomeres. This suggests that it's no coincidence that regular exercisers have longer telomeres—and though this effect is invisible to the naked eye, it may prove to be fitness's most important benefit.

CHEAT SHEET: EXERCISE AND AGING

- Starting in your mid-30s, you lose 1 to 2 percent of your muscle mass each year and about 9 percent of your aerobic fitness per decade—but regular exercise slows this decline dramatically.
- Long-term studies find that runners get osteoarthritis at a lower rate than non-runners, contradicting the common belief that running wears down your knees.
- Successful masters athletes train consistently without long breaks, focus their workouts on the most essential elements, and take extra recovery time to avoid injuries.
- Endurance declines more sharply than speed as you age. Steady training may prevent your rate of decline from accelerating.
- Declining motivation may be as important as aging bodies in explaining why older athletes slow down. Ensuring that your family and friends are supportive helps maintain positive social pressure.
- Aerobics-style exercise in water can reduce the impact on joints and lower the risk of falls. The exercise benefits are similar to dry land, though your heart rate will be lower due to water pressure.
- Activities that build muscle (like strength training) or provide jarring impacts (like running or basketball) are better for building strong bones than cycling, swimming, or elliptical training.
- Exercise slows down the cellular aging process in which the caps on the end of your DNA (known as telomeres) get shorter.

9: Weight Management

THE "SECRET" TO EASY WEIGHT LOSS is well-known—so well-known, in fact, that everyone has one. It's eating less fat . . . or fewer carbs. It's exercising at low "fat-burning" intensity . . . or at maximum intensity. It's consuming fewer calories . . . or burning more calories. It's lifting weights . . . or doing cardio. The contradictions are endless, and the real message is that there's no single tactic that makes losing weight easy for everyone.

But that doesn't mean we don't know anything about weight loss. On the contrary, researchers are teasing apart the complex links between diet, physical activity, hormones, and fat storage. Understanding how different types of exercise and eating patterns affect the body will help you plot a weight-loss strategy tailored to your needs and avoid common misconceptions like the myth of the "fat-burning" zone. More importantly, it's now clear that being thin and being healthy aren't always the same thing—so the success of your exercise regime should be measured by aerobic fitness, not the bathroom scale.

Is it possible to be fat and healthy at the same time?

Fat is bad and obesity is an epidemic—that's the message we hear on a daily basis. That's why the unexpected results of a 2009 study published in the journal *Obesity* created a mini-sensation. A team of Canadian and American researchers used data from

Statistics Canada's National Population Health Survey to follow 11,326 adults for 12 years. They found that subjects who were overweight (body mass index of 25 to 30) were 17 percent less likely to die during the study period than those of normal weight (BMI of 18.5 to 25).

Does this mean we've been wrong all along about the links between weight and health? That our mantra should be, as newspaper headlines put it, "Get fat, live longer?" In truth, the results weren't surprising to obesity researchers, joining a growing pile of evidence that body weight is not the absolute indicator of health we once thought. But don't quit the gym yet. It turns out that physical fitness is a far better barometer of your long-term health than weight is—and that holds true even for thin but inactive people who thought their fabulous metabolism meant they didn't need to exercise at all.

Steven Blair, a professor at the University of South Carolina, has performed a series of studies dating back to 1994 that try to distinguish between obesity and physical inactivity as causes of health problems. "When we look at obesity and properly adjust for fitness, the obesity risk goes away," he says. "It just disappears." In fact, he says, obese people who are physically fit are half as likely to die as people of normal weight who don't exercise.

This message is particularly crucial for people who start exercising and soon get frustrated—and perhaps quit—because they don't succeed in losing weight. As long as they're meeting basic exercise goals such as half an hour of moderate to vigorous exercise five times a week, Blair says, they're gaining important health benefits no matter what the scale says. Seen in this light, the Statscan results are less shocking—in fact, they closely mirror the results of a similar U.S. study from 2005, which also found that those carrying a few extra pounds into old age lived longer.

"As you age, you start to lose weight and become frail," notes one of the Statscan study's authors, David Feeny of Kaiser Permanente Center for Health Research in Portland. For the elderly, among whom most of the deaths in the study occurred, a few extra pounds may provide a margin of error to help them through the illnesses and accidents that become common at that age. In addition, a more vigilant health care system that watches for warning signs such as high blood pressure may have actually succeeded in lowering the penalty for being obese over the past few decades, Feeny says.

The study doesn't give obesity a free pass—those with a BMI above 35 were 36 percent more likely to die during the study than those of "normal" weight. But researchers also point out that BMI isn't the most effective way to measure risky fat build-up. "It's most useful in population studies," says Travis Saunders, a researcher at the Children's Hospital of Eastern Ontario Research Institute in Ottawa whose blog Obesity Panacea covers the latest findings in obesity research. "But if you try to apply it to individuals, it doesn't work." That's because *where* you store fat is as important as how much you have. Fat in the abdominal region, particularly the visceral fat that accumulates between internal organs rather than fat stored just beneath the skin, is particularly problematic. In contrast, Saunders says, fat on the hips, buttocks, and lower body appears to be less of a concern. For that reason, many doctors now measure waist circumference as a proxy for visceral fat. Ideally, men should be less than 40 inches (102 centimeters) and women should be less than 35 inches (88 centimeters).

It's also worth remembering that (strange as it may sound) death isn't everything. The average lifespan is now long enough that conditions like heart disease, hypertension,

BODY MASS INDEX

To calculate your body mass index (BMI), divide your mass in kilograms by the square of your height in meters. For example, a man who is 1.75 meters tall and weighs 75 kilograms would have a BMI of $75 / (1.75)^2 = 24.5$. (Alternatively, multiply your weight in pounds by 703 and divide by the square of your height in inches.) A BMI between 18.5 and 25 is considered normal weight; below 18.5 is underweight and above 25 is overweight, while above 30 is defined as obese.

Sample BMIs				
	125 lbs	175 lbs	200 lbs	225 lbs
5'0"	24.4	34.2	39.1	43.9
5'6"	20.2	28.2	32.3	36.3
6'0"	17.0	23.7	27.1	30.5

and diabetes—all of which are strongly linked to being overweight—can have a serious impact on quality of life in your final years. The message of the Statscan research, ultimately, is not that weight is irrelevant but that your focus should be on the ongoing process of living healthily, rather than the potentially misleading endpoint of reaching a certain weight.

Is weight loss simply the difference between "calories in" and "calories out"?

In theory, managing your weight is simplicity itself. If you take in more energy (in the form of food) than you burn (through physical activity and metabolic processes), those extra calories are stored as excess weight. If you burn more than you eat, you lose weight. Calories in minus calories out. From a physicist's perspective, this is inarguably true, since nature forbids you to either create or destroy energy. In practice, though, it's a little more complicated—because the "calories out" part of the equation doesn't behave as we expect.

Consider what would happen if you added one 60-calorie chocolate-chip cookie to your daily diet. Since a pound of fat contains 3,500 calories, simple math suggests that you'd pack on about half a pound a month, or six pounds a year, for the rest of your life. But, as a 2010 paper in the *Journal of the American Medical Association* explains, this isn't what happens. As you begin to gain weight, your body has to spend metabolic energy repairing, replacing, and supporting the cells in this new tissue. You start burning more calories without any change in physical activity. As a result, the weight gain slows down and eventually levels off after a few years at a total of six pounds—even if you keep eating that extra cookie for the rest of your life.

Unfortunately, the opposite happens if you start eating 60 fewer calories per day. Initially, you'll lose weight. But now the body doesn't have to expend any energy maintaining the lost tissue, so you burn fewer calories, and the weight loss levels off. If you now revert to your normal diet—which is what people usually do once they achieve their weight-loss goals—you'll simply regain the weight.

It's not just the weight you gain or lose that conspires to keep your body weight constant. For more than a decade, researchers at Columbia University have been performing painstaking experiments in which volunteers check into a controlled hospital setting for months at a time. They're fed only a liquid diet (40 percent of the calories come from corn oil, 45 percent from glucose, and 15 percent from the protein casein) so that their daily caloric needs can be computed exactly. With both obese and normal-weight subjects, the researchers control food to reduce or increase body weight by 10 percent and observe the metabolic consequences.

The most recent study, published in 2010 in the *American Journal of Physiology*, tested how the subjects responded to physical activity. When they lost weight, their muscles became about 15 percent more efficient—not simply because they were carrying less weight around, but because of changes in the ratio of enzymes responsible for burning fat or carbohydrate as fuel. While greater efficiency sounds like a good thing, it means they were burning fewer calories, making it harder to maintain the low weight. In contrast, their muscles became 25 percent less efficient when they gained weight, once again pushing them back toward their starting weight.

These findings provide a sober reminder of how hard it is to slim down once you've gained weight. But there are also some practical lessons we can draw from the Columbia research. The biggest change in muscular efficiency was observed at the lowest levels of exercise intensity, equivalent to the activities of day-to-day life as opposed to exercise. As a result, the researchers suggest, "the weight-reduced individual might 'escape' this increased efficiency by altering the intensity of exercise." Exercising harder rather than longer might be the most effective strategy for keeping weight off.

To lose weight, is it better to eat less or exercise more?

A calorie is a calorie is a calorie, according to the old-school, keep-it-simple school of nutritional thinking. But as we've seen, good health is a little more complicated than just matching the calories you eat to the calories you burn. A study published in 2010 in the journal *Medicine & Science in Sports & Exercise* suggests that your body can tell the difference between a calorie burned through exercise and a calorie avoided through dieting—and both types turn out to be important.

Researchers at Louisiana State University recruited 36 moderately overweight volunteers and divided them into three groups. One group was the control and stayed exactly the same throughout the six-month study. A second group cut their calorie intake by 25 percent, while the third group cut calories by 12.5 percent and increased their calories burned by an equivalent amount through physical activity. That meant both intervention groups had the same total "calorie deficit"—one through diet alone, and the other through a 50-50 mix of diet and exercise.

As expected, the two intervention groups lost exactly the same amount of weight—a fairly impressive 10 percent of their starting weight. They also lost about 25 percent of their total body fat and 25 percent of their abdominal fat, again with no difference between the two groups. This confirms that the amount of weight you lose is a function of calorie deficit, whether you create the deficit through diet or exercise. But when the researchers took a closer look, they found some important differences between the two groups. Only the diet-plus-exercise group had significant improvements in insulin sensitivity, LDL cholesterol, and diastolic blood pressure—crucial risk factors for heart disease and diabetes, but changes you can't measure by looking in the mirror or stepping on a scale.

This research sheds new light on an ongoing "fitness v. fatness" debate. While some researchers believe that weight and body mass index are the simplest measure of cardiovascular risk, others—most prominently Steven Blair of the University of South Carolina—argue that aerobic fitness is more important than body shape. It's difficult to tease apart the effects because, in general, those who are fattest are the least fit and vice versa. The Louisiana State results support Blair's position by showing that there are some key health benefits that you can't get just by being skinny: you have to exercise too.

But that's not the final word. Two other risk factors measured in the Louisiana State study—systolic blood pressure and HDL cholesterol—remained identical between the two experiment groups. That suggests that they depend on weight rather than aerobic fitness, so that "fat but fit" people may still be missing out on some health benefits. The challenging but inevitable conclusion is that both diet and exercise are important to optimize your health, and you can't ignore either of them.

How can I take advantage of the "fat-burning" zone?

If you're looking to shed fat, wouldn't it be nice if you could choose to selectively burn fat instead of wasting your time burning carbohydrates? That's the idea behind the famous "fat-burning" zone touted by exercise equipment makers and fitness gurus. Not only that, but the key to staying in the fat-burning zone is not to push too hard—a welcome message indeed. Unfortunately, there are flaws of both logic and physiology behind these claims.

Let's start with the kernel of truth. You do indeed burn a mix of fat and carbohydrate when you exercise, and the exact proportion varies with your exercise intensity. When you go for a leisurely walk, you might burn 85 percent fat and 15 percent carbohydrate. If you pick up the pace to a jog, you'll start burning more carbohydrate. The harder go you, the more carbohydrate you burn, until you're burning about 70 percent carbohydrate and 30 percent fat at the highest intensities. The cross-over point, where you get half your energy from each source, occurs at about 60 percent of your maximum intensity (though it can vary considerably from person to person and increases as you get fitter).

On the surface, it sounds like low-intensity is indeed the way to go for fat-burning. But this ignores the question of total calories burned. If you go for a leisurely walk and burn 100 calories, it's true that 85 of them will be provided by fat. But it's a lot better to go for a moderate run and burn 500 calories in the same amount of time, with 250 from fat. (Exercise at the highest intensities has many benefits for health and fitness, but it's hard to sustain for long periods of time—so from a fat-burning perspective, moderate-intensity exercise is the most effective.)

But there's a more fundamental problem with the fat-burning idea, relating to how your body recovers after physical activity. If you burn mostly carbohydrates during a workout, the calories you consume in the hours after the workout will be used to replenish your depleted carbohydrate stores. If you manage to rely more on fat during the workout, on the other hand, your carbohydrate stores will remain full. As a result, whatever calories you consume afterwards will be stored directly as fat, undoing your fat-burning efforts.

This phenomenon was demonstrated in 2010 by researchers at the Garvan Institute of Medical Research in Australia. In mice that were genetically altered to burn fat instead of carbohydrate, unburned carbohydrates were simply converted into fat for storage. The total "energy balance" remained unchanged, and the mice gained or lost the same amount of weight as normal mice under the same conditions. The results serve as a warning not to waste your time and money on pills that claim they enhance fat-burning, according to Greg Cooney, the study's senior author. "Our data urges a correction in people's concept of a magic bullet—something that will miraculously make them thin while they sit on the couch watching television," he said.

So far, the only technique that reliably boosts the proportion of fat you burn is—you guessed it—exercise. After a few months of training, studies have found that you will indeed burn more fat at a given level of intensity than you did when you were out of shape. Of course, while that may help you stay fueled in the late stages of running a marathon, it doesn't really matter for weight-loss purposes. The only thing that counts is how many calories you burned to get there.

Won't exercise make me eat more and gain weight?

In 2009, *Time* magazine ran a cover story called "Why Exercise Won't Make You Thin," in which journalist John Cloud described his unsuccessful attempts to lose weight through exercise. Cloud's central theme, supported by his own fondness for post-workout blueberry bars and a rather selective sampling of research, was that exercise actually makes you hungrier. As a result, he warned, "fiery spurts of vigorous exercise could lead to weight gain." The twisted logic that led to this conclusion was widely condemned by obesity researchers—but it did raise an important question. After all, it's undeniably true that many people exercise diligently without losing weight. And exercise does make you hungrier.

Some simple math illustrates what you're up against. Let's say you go out and bike six miles in about half an hour, then chug a typical recovery shake. You've burned about 280 calories and immediately downed 270 calories—so you haven't accomplished much. The number of calories burned through casual exercise almost always corresponds to a surprisingly small chunk of food. That means dropping weight is not an easy process. But there's no evidence to suggest that exercise actually causes you to gain weight.

It's true that increasing your physical activity levels can make you feel hungrier, but the same is true of eating less. Your body will respond to any change that results in you taking in fewer calories than you burn with a series of physiological and behavioral tactics that conspire to keep you at your current weight. That's why almost none of the weight-loss interventions that have been tested in clinical trials achieve losses that the majority of participants sustain beyond a few years. It's not just exercising to lose weight that's hard—it's losing weight by any means.

Of course, there's no debate that elite athletes drop pounds and keep them off through exercise. In fact, for long-distance runners, swimmers, and Tour de France cyclists, eating enough to meet their caloric needs is a constant challenge. So it's clear that exercise really can help you lose weight—the only question is how much. A recent Harvard University study offers some clues. Researchers followed 34,000 middle-aged women for 13 years, monitoring their diet, exercise, and weight and reporting the results in the *Journal of the American Medical Association* in 2010. Just 13 percent of these women (who were eating "typical" American diets with no intervention) managed to avoid significant weight gain throughout the study, and these women averaged a full hour of moderate exercise every day. Anything less was unsuccessful. That's a lot of exercise—unless you compare it to the daily lives of our ancestors who didn't spend most of the day sitting at desks or in cars.

In a sense, Cloud's article was a wake-up call to anyone who thought that heading to the gym for half an hour a few times a week would, on its own, transform their bodies. You also have to pay attention to what you eat, both immediately after your workout and throughout the rest of the day. But the article's most serious sin was underplaying the other benefits of exercise, from cardiovascular health to stress relief, that accumulate even if your weight isn't changing. Exercise—and particularly "fiery spurts of vigorous exercise"—is the most powerful force for good health that we know of. And it won't make you gain weight.

Can I lose weight while gaining (or maintaining) muscle?

In a perfect world, you'd be able to exercise and eat in a way to make your muscles grow bigger while your fat stores shrink. This is possible in some cases—but for most of us, a more realistic goal is to lose weight by dropping fat without losing useful muscle mass. This is particularly important for athletes who compete in weight classes, like wrestlers, who don't want to weaken themselves while trying to "make weight." Researchers have been studying the problem for the past few years, and they now believe that one of the keys is making sure you're getting enough protein.

Protein gets a lot of hype for its weight-loss potential. For one thing, foods containing protein make you feel more full, so you're likely to consume fewer calories overall. Some research suggests that protein helps maintain levels of a hormone called triiodothyronin, which counteracts the tendency of your resting metabolism to slow down when you lose weight. Protein is also a relatively inefficient fuel: your body has to burn 25 percent of the available energy just to convert it to a usable form. These factors all sound great, but it's not yet clear that they make any appreciable difference in your attempt to lose weight and maintain muscle. There's stronger evidence for the role of leucine, one of the essential amino acids provided by dietary protein, in stimulating your body to create more muscle protein.

In practice, several studies have found that higher protein diets help overweight or obese subjects lose more fat while retaining more muscle compared with diets with less protein. But it's important to clarify what "higher protein" means in this context. For example, the diets in a 2005 study at the University of Illinois contained either 0.8 grams of protein per kilogram of body weight (g/kg/day) or 1.6 g/kg/day. In this case, the "high-protein" diet actually corresponded to the amount of protein in a "typical" North American diet, which is about 1.6 g/kg/day— equivalent to about three-quarters of a pound of chicken breast for someone who weighs 150 pounds. So the message isn't to eat more protein; it's just to avoid slashing your protein intake if you're cutting calories.

Athletes in training face somewhat different challenges. They have less fat to start with, which means any weight loss is more likely to come from muscle. But the same basic pattern holds true, according to a 2010 study at the University of Birmingham. Researchers found that a diet containing 2.3 g/kg/day of protein (35 percent of calories) helped athletes maintain their muscle mass far better than a diet containing 1.0 g/kg/day (15 percent of calories). They lost roughly the same amount of fat on both diets.

Another key factor that athletes need to consider before ramping up their protein intake is that carbohydrates are the most important fuel for extended bouts of exercise. A study in New Zealand found that even a single week on a high-protein diet with 3.3 g/kg/day of protein was enough to slow endurance cyclists by 20 percent on a two-hour time trial, compared with a high-carbohydrate diet with just 1.3 g/kg/day of protein. This suggests that you shouldn't emphasize protein at the expense of carbohydrate, which leaves fat as the only remaining candidate for reduction.

Ultimately, the message is balance. Too little protein, and you'll lose muscle mass (assuming you're also eating fewer calories overall than you're burning). Too much protein at the expense of carbohydrates, and you'll suffer at the gym. As long as you're realistic about your weight-loss goals, aiming for no more than a pound a week, you can get there without dramatic deviations from a normal, balanced diet—one that doesn't go heavy on the protein.

Is lifting weights better than cardio for weight loss?

Conventional wisdom says that aerobic exercise (combined with cutting calories) is the best way to lose weight. But millions of people have logged hours on elliptical machines and stationary bikes without dropping any pounds. This means that either (a) losing weight takes a lot more effort than most people expect, or (b) we've been misled and there's a much better way to lose weight. Strength training is often proposed as that "better way"—though the evidence strongly suggests the real answer is (a).

In a head-to-head match-up of an aerobic workout versus a strength workout, there's no dispute that you'd burn more calories in the aerobic workout (assuming that intensity and duration are held roughly equal). It's the calories you burn during the rest of the day that might tilt the field in favor of weights. A classic study published in 1977 showed that the gradual decline with age of your resting metabolic rate—the calories you burn just to stay alive, even when you're sleeping—is due almost entirely to the loss of muscle mass that begins in your mid-30s and continues inexorably for the rest of your life. Pumping iron slows the loss of muscle, or even adds new muscle, which keeps your metabolism ticking a little more quickly.

Strength training also stimulates your body to burn more fat instead of carbohydrate as fuel—though it's not clear that burning more fat actually reduces the amount of fat you store in the long term (see p. 190). There's also the simple fact that if you're strong and healthy, you're more likely to move around, lift things, climb stairs, and otherwise burn calories in the course of your day-to-day life. It was these factors that convinced the American College of Sports Medicine, in 2009, to acknowledge the possibility that strength training might contribute to weight loss, reversing an earlier official stand. They're now taking more of a wait-and-see approach—there's no experimental evidence that these factors make any significant difference, but they at least sound plausible.

There's no shortage of studies on strength training and weight loss. A typical example, published in the *American Journal of Clinical Nutrition* in 2007, monitored 164 overweight, middle-aged women for two years. Half of them lifted weights twice a week, the other half were simply given brochures recommending aerobic exercise. The weights group gained about three pounds, including a 7 percent increase in dangerous abdominal fat; the control group gained 4.4 pounds, with a 21 percent spike in abdominal fat. This is clear evidence that strength training is good for you—but not that it's better than aerobic exercise for weight loss.

The most positive results, not surprisingly, come from studies that combine aerobic and strength exercise. A Korean study that pitted a six-days-a-week aerobic training program against three days each of aerobic and strength training found that the combined program produced the best results for decreasing surface and abdominal fat, as well as increasing muscle mass. There's no doubt that strength training has innumerable benefits—including, possibly, boosting your metabolism and fat-burning abilities. But for real-world health purposes, it works best in combination with aerobic exercise.

Will I burn more calories commuting by bike or on foot?

Most commuters strive to be as efficient as possible. But to get the best workout—specifically, to burn the most calories—you're better off being inefficient. Biking three miles to work could burn 130 calories, while walking the same distance would burn 225. Of course, the bike ride would take 15 minutes, while the walk would take close to an hour. If you add a scenic detour to the bike ride so that it takes an hour, you'd burn over 500 calories. (Those numbers are for a 155-pound person biking at 13 miles per hour or walking at three to four miles per hour, both "moderate" paces.)

You'll need to balance these different types of efficiency—time spent versus energy burned—in order to choose the right mode of transport for your commute. Your decision will also depend on logistical factors like the length of the commute, where there are good bike paths, and whether your workplace has showers. If you're lucky, you'll be able to alternate between different options.

- **BIKING:** The dominant factor in outdoor biking is air resistance, which gets increasingly important the faster you go. It accounts for 90 percent of the resistance you feel at racing speeds above 18 miles per hour. This means that if you bike too slowly it will be even less taxing than a brisk walk. But it also means that cranking up the cadence can quickly make it harder. Hills and frequent stops can also make the bike commute more of a workout. Because of the trade-off between air resistance and gravity, the most time-efficient strategy is to push a bit harder on the up-hills (when your speed is already reduced) and recover on the downhills. Biking is the most practical commuting option for any distance beyond a few miles. But it takes conscious effort (along with, perhaps, a shower at work and a detour on the way home) to turn it into a really good workout.

- **WALKING:** Each of us has an optimal walking speed that feels comfortable and burns the fewest calories per mile. Walking significantly faster than that optimal speed can actually be as fast and calorie-intensive as a very slow run, but it quickly becomes uncomfortable to sustain. That's why, with all due respect, race-walkers are so funny to watch. It's important to bear in mind the differences between a brisk walk and a leisurely stroll. You're burning more than three times as many calories at five miles per hour than you are at two miles per hour. Unlike a gym workout, the length of your commute is determined by distance rather than time. Chalk that up as an advantage for walking, which will make your workout commute the longest—but make it a brisk walk that requires effort.

- **RUNNING:** There's a long-standing misconception that propelling yourself on foot from point A to point B takes a

COMMUTING CALORIES

For a given distance, walking burns more calories than biking. For a given time, it's the other way around. Either way, running is the fastest way to burn calories. These calculations are for someone weighing 145 pounds, on flat terrain.

WALKING   RUNNING   BIKING

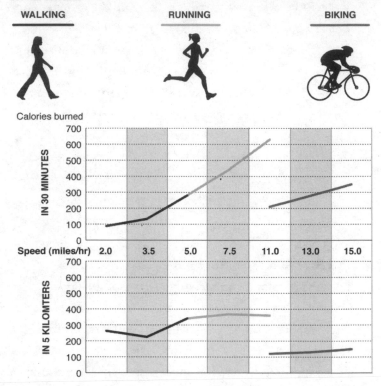

set number of calories no matter how fast you move. That idea was finally debunked in a 2004 study in the journal *Medicine & Science in Sports & Exercise,* which showed that running at 10 minutes per mile burns about twice as many net calories as walking at 20 minutes per mile. The difference comes in part from the up-and-down motion of the running stride. Running offers commuters the perfect mix of a vigorous but time-efficient commute, but it comes with its own logistical challenges. Not only do you have to shower, but you need to organize (and perhaps schlep in a backpack) clothing and possibly lunch, as well as decide how to get home.

Can I control hunger by manipulating my appetite hormones?

In recent years, scientists have identified a set of hormones that control eating behavior. For example, rising levels of ghrelin signal that it's time to start eating, while rising levels of leptin tell you that you're full. A German study in 2008 showed that even a single night of shortened sleep raises levels of ghrelin, explaining why you often crave snacks when you're tired Similarly, just two nights of short sleep cause a drop in the fullness hormone, leptin. The same is true if you consistently get even an hour or two less sleep than you need. So it's not surprising that studies have found a direct correlation between how many hours of sleep you get and how thin you are.

Eating patterns can also influence these hormones. In a 2010 study, Greek researchers had volunteers eat identical bowls of ice cream in either 5 minutes or 30 minutes. (For the slow eaters, the researchers divided the ice cream into seven equal parts and fed it to the volunteers every five minutes, so that it wasn't melted by the end of the half-hour.) In this case, although there was no difference in ghrelin levels, the researchers did observe significantly higher levels of peptide YY and glucagon-like peptide (two gut hormones that signal fullness) in the slow-eating group. This group also reported feeling more full.

Another long-standing staple of dietary wisdom is that you should eat frequently rather than cramming all your calories into three big meals. The idea is to prevent large hunger swings by keeping the levels of appetite-determining hormones in your gut relatively constant. But studies over the past half-century have reached conflicting conclusions about whether this actually works. Most recently, researchers at the University of Ottawa put 16 obese volunteers on diets with identical caloric deficits for eight weeks, publishing the results in the *British Journal of Nutrition* in 2010. Half of them ate three meals a day, while the other half ate three meals plus three snacks, with the total intake tailored so that each subject was burning 700 more calories than he or she consumed each day. By the end of the study, the subjects had lost an average of 4.7 percent of their starting weight, but there was no difference between the two groups. The researchers also measured the hourly fluctuations in ghrelin and peptide YY but didn't find any significant differences between the two groups.

The University of Ottawa study suggests that snacking doesn't have any miraculous appetite-reducing effects, but the subjects weren't regular exercisers. If you do work out regularly, the rules are slightly different. It's a good idea to eat something very soon after exercising (see p. 211)—it could be either a meal or a snack, depending on your schedule. This will help you recover from the workout, and there's some evidence that it might help you avoid overeating later.

Will sitting too long at work counteract all my fitness gains?

You'd think that spending an hour a day sweating at the gym would be enough to guarantee good health. But a 2010 study in the *American Journal of Epidemiology* added to growing evidence that what you do during the rest of the day also makes a difference. The researchers followed 123,000 people for 13 years and found that men and women who spent more than six hours per day sitting down were 18 and 37 percent, respectively, more likely to die during the study than those who sat fewer than three hours per day. What's most surprising is that these risks were completely unrelated to how much exercise the subjects reported getting.

Scientists aren't yet sure why spending long periods of time sitting down should cause health problems, but they view it as a sign that the low-intensity activity associated with simply walking around and doing everyday chores makes an important contribution to health. Research at the University of Massachusetts Amherst suggests that it can also make a key contribution to weight loss, since it's low-key enough that it doesn't spark hunger to compensate for the calories burned. A forthcoming Amherst study compared a group of volunteers who sat all day (they even used wheelchairs to visit the bathroom) with a group that didn't sit down at all. Preliminary results show that the difference in energy expenditure was hundreds of calories—but the level of appetite hormones and reported hunger in the two groups remained identical.

Of course, for people who work in office settings, walking around all day isn't really an option. Some experts recommend scheduling regular "micro-breaks" every 30 to 60 minutes, in which you stand up, stretch, and walk away from your desk for a few minutes (preferably not to the fridge). Free downloadable programs like Workrave (www.workrave.org) provide periodic warnings to remind you when to take these breaks.

Another low-intensity calorie-burning option is to replace your desk chair and use an exercise ball instead or even switch to a standing desk. A 2008 study by researchers at the University of Buffalo found that either sitting on an exercise ball or standing resulted in an extra 4.1 calories burned per hour compared to sitting in a regular office chair. Best of all, the typing rate of the subjects in the study was the same in all three cases. If you try either of these options, don't go "cold turkey"—start with no more than a few hours a day. Also, be alert for signs of lower-back pain when sitting on the exercise ball, since the lack of support could expose existing weaknesses in stabilizing muscles.

None of this suggests that more vigorous exercise isn't also important. For example, a 2010 University of Western Ontario study that compared low- and high-intensity activity found that easier exercise acts primarily on your heart, while harder exercise acts on your muscles. You need both your heart and muscles to be healthy, so don't try to get away with just one of the options. In fact—and this is a message that applies in almost every aspect of fitness, diet, and health—the very best approach you can take to choosing your exercise intensity is to avoid doing the same thing every day.

CHEAT SHEET: WEIGHT MANAGEMENT

- Obese people who are physically fit are half as likely to die as thin, sedentary people. Aerobic fitness may be a better measure of health than body-mass index.
- When you lose weight, your muscles become more efficient and your metabolism slows down in an attempt to regain the weight. More vigorous exercise may avoid this "efficiency trap."
- Cutting food intake or increasing exercise by the same number of calories produces the same amount of weight loss; however, some improvements in blood pressure, cholesterol, and other factors require exercise.
- Low-intensity exercise burns a higher proportion of fat than high-intensity exercise, but fewer calories overall. Weight loss depends on overall caloric deficit, because your body converts unburned carbohydrates to fat for storage.
- Losing weight through exercise alone is very challenging: middle-aged women had to exercise for an hour a day just to avoid gaining weight in a recent long-term study.
- If you're consuming a low-calorie diet to lose weight, increasing the amount of protein you consume to 35 percent of calories helps avoid muscle loss.
- Aerobic exercise burns the most calories, but building muscle through strength training helps keep your metabolism high. Combine both approaches for the best results.
- To cover a given distance, running burns more calories than walking, which burns more calories than cycling. In a set amount of time, cycling burns more calories than walking.
- Eating slowly and getting enough sleep both help to control the hormones governing your appetite. Snacking between meals doesn't appear to have a major effect on those hormones.
- Sitting all day at work can negatively affect your health, no matter how many hours you clock at the gym. Make sure to schedule regular breaks from your desk to stand up and walk around or stretch.

10: Nutrition and Hydration

WHEN THE GREEK RUNNER SPIRIDON LOUIS won the first Olympic marathon in 1896, he reportedly stopped along the route to consume wine, milk, beer, orange juice, and an Easter egg. Sports nutrition has come a long way since then, but figuring out what and when to eat and drink remains challenging. There's plenty of competing information on how to fuel up before and during exercise, how to refuel afterwards, and whether dehydration is really such a bad thing. No discussion of sports nutrition is complete without a look at some of the powders and pills that claim to (legally) enhance physical performance. Despite hundreds of scientific studies, there's little evidence to back up the claims made by most supplements. But there is some interesting research emerging regarding supplements like vitamin D and probiotics, which makes them worthy of a closer look.

Should I carbo-load by eating pasta the night before a competition?

The origins of the pre-race spaghetti dinner go back to pioneering Scandinavian studies in the 1960s. Researchers found that if you started depleting your carbohydrate stores a week before competition by exercising hard and eating only fat and protein, your body would "super-compensate" by storing extra carbohydrate in your muscles when you carbo-loaded in the final few

days. This led athletes to adopt a difficult and often unpleasant week-long carbohydrate depletion-loading regime before
competitions.

There was one key problem with the initial studies: they
used untrained subjects. Later researchers showed that trained
subjects depleted their carbohydrate stores through the daily
act of training and gained no additional benefit from adding a
depletion phase. An Australian study in 2002 took things one
step farther, showing that you can maximize your carbohydrate
stores by eating 10 grams of carbohydrate per kilogram of body
weight for just one day, as long as you don't exercise vigorously
that day. Continuing on a high-carbohydrate diet for another
two days did nothing to further enhance carbohydrate stores in
the Australian study.

It's important to note that 10 grams of carbohydrate per
kilogram is a very large amount—far more than you'll get from
a simple pasta meal. In fact, researchers have found that most
athletes fail to fully max out their carbohydrate stores. If you're
a 70-kilogram (155-pound) athlete trying to pack in 700 grams
of carbohydrate, you'd need to eat about 10 plates of spaghetti.
Even over the course of a full day, it's difficult to eat this much
without supplementing your regular meals with sports drinks
and other concentrated sources of carbs.

Having a full tank of carbohydrates won't increase your peak
running, cycling, or swimming speed, but it should allow you to
maintain your pace for a little longer before your carbohydrate
stores are depleted and you "hit the wall." As a result, there's
no particular benefit to carbo-loading for shorter events where
you don't have time to completely deplete your body's stores.
For exercise lasting less than about 90 minutes—researchers disagree about the exact threshold—you don't need to carbo-load.

Even for longer races like marathons, the evidence isn't as clear as once thought. In the studies that showed carbo-loading benefits, subjects often weren't allowed to consume any carbo-hydrates during the exercise trial. In real life, though, you're free to drink sports drinks or swallow gels during marathons or long tennis matches. When you're allowed to top up your stores like this during the actual event, the evidence that carbo-loading helps is much weaker. And there's a potential down-side: when you store carbohydrates, your body stores water along with it, meaning that a successful bout of carbo-loading can add several pounds to your competition weight. How you should balance these pros and cons depends on your personal experience—whether your stomach can handle ingesting carbo-hydrates during a competition, whether you've "hit the wall" in previous competitions, and so on.

There is one other factor that a pasta dinner—or any day-before loading, for that matter—can't handle. While you sleep, about half the carbohydrates stored in your liver as glycogen will be consumed to fuel your nervous system. These liver glycogen stores help maintain normal blood sugar levels and fuel crucial organs like your brain, which can't access all the glycogen stored in your muscles. For that reason, top endurance athletes make sure they get up several hours before competition (no matter how early that is!) and eat some easily digested carbohydrates like a banana, oatmeal, or a bagel to top up their liver glycogen.

What should I eat to avoid stomach problems during exercise?

Your stomach gets 10 to 20 times bigger as it goes from empty to full, ultimately holding about four cups of food and drink on aver-age. This is a neat trick, but can lead to problems during exercise if you're not careful. Studies have found that as many as half of people who do sustained aerobic exercise suffer from gastrointestinal problems like cramping, nausea, or diarrhea. Others find that mistimed meals leave them dizzy or short of energy, or cause a stitch (see p. 70). By following a few simple guidelines, you can reduce the chances that you'll run into these problems.

After you swallow a mouthful of food, it typically takes an hour or two to move through your stomach into your colon. (It will be 24 to 72 hours before it finally exits the body.) For this reason, it's best to allow three to four hours between a meal and a hard workout. Researchers have found that intense training can speed up this "orocecal" (mouth-to-colon) transit time, probably because your stomach gets used to processing the large amounts of food necessary to keep up training. In one study at Indiana University, collegiate swimmers and distance runners consuming 4,000 to 5,000 calories a day had orocecal transit times as low as half an hour, while sedentary controls consuming less than 2,000 calories a day took as long as three hours. (These times represent the speed for the first mouthful; it may take much longer to process a full meal.) This means that as you get fitter, you'll get better at moving food through your system while still absorbing as many nutrients.

If you still have food in your stomach when you start exercise, the digestion process will slow down as blood is diverted away from the gut to your working muscles. In sports like running, the constant up-and-down jostling of your stomach and its contents may contribute to the chance of a cramp. Stress can also slow digestion, so you need to allow extra time for food to clear the stomach if you're nervous before a race.

Additionally, your choice of foods can make a big difference. Dietary fiber slows down digestion and also increases bulk in your colon by drawing in water—so you're better off with white bread than whole-wheat for a pre-workout snack if you've been having GI problems. Foods high in fat also take longer to move through the stomach. Unfortunately, pre-workout carbohydrates come with problems of their own: some people experience an effect called "rebound hypoglycemia" after eating carbohydrates in the hour before exercise, resulting in dizziness, weakness, and sometimes nausea after 15 or 20 minutes of exercise. This happens because simple carbohydrates trigger a rise in insulin to reduce levels of blood sugar. Exercise also reduces blood sugar, so if you combine the two back to back, your blood sugar levels drop too low and you get light-headed.

One way to avoid rebound hypoglycemia is to abstain from carbohydrates for the hour before your workout. Or you can take the opposite approach: if you eat carbohydrates in the last five minutes before you start, your insulin levels don't have time to spike, thus avoiding the problem. It also helps to stick to foods with a moderate or low glycemic index, which means they cause a slower rise in blood sugar. A banana, a bowl of oatmeal, or a piece of whole-wheat bread with peanut butter are good examples of foods with a moderate glycemic index.

Finding the approach that works best for you requires trial and error—and it may involve avoiding foods that don't cause you any trouble under normal circumstances. For reasons that still aren't entirely clear to researchers, long or intense bouts of exercise seem to make the digestive system hypersensitive even to very minor food intolerances. With careful experimentation, though, you should be able to find a few simple meals that your stomach can tolerate even under the toughest conditions.

What should I eat and drink to refuel after working out?
It's always nice when science tells you what you want to hear.
That's why several studies in the past few years touting low-fat
chocolate milk as a perfect post-workout elixir have been greeted
so enthusiastically. Chocolate milk is convenient, cheap, and
tasty, so what's not to like? But you have to be cautious when you
take research done on competitive athletes and try to apply the
results to casual exercisers. The basic principles are the same, but
new research—and common sense—suggests that those whose
main goal is to lose weight should chug milk with caution.

Post-exercise nutrition has two primary goals. First, you want
to recharge your body's depleted energy stores so that you're fresh
and recovered before the next workout—whether that's coming
later the same day or later in the week. Second, you're maximiz-
ing your fitness gains by providing the raw materials your body
needs to synthesize the contractile proteins that increase strength
and the mitochondrial proteins that boost endurance. "It's a con-
tinuum between short-term recovery and long-term adaptation,"
says Trent Stellingwerff, a Canadian scientist in the performance
nutrition group at the Nestlé Research Center in Switzerland.

The key factors to consider are when and what you eat. For
the first half-hour after exercise, the body is processing nutrients to repair itself at a dramatically elevated rate. After about
two hours, this "window" is closed and the opportunity for any
accelerated recovery is lost. This is where high-tech recovery
bars and drinks can be useful, since they're easy to have on
hand immediately after you finish exercising—though it's also
easy to pack a sandwich in your workout bag.

In the past, conventional wisdom held that weightlifters
should ingest protein to build muscle, while endurance athletes
should focus on carbohydrates. Now researchers agree that both

macronutrients are important no matter what type of exercise you're doing, Stellingwerff says. Recent studies suggest that you should aim to consume about one gram of carbohydrate and 0.3 grams of protein per kilogram of body weight during the first hour or two after a typical cardio workout.

Some sports drinks tout very specific carb-to-protein ratios that companies say will optimize recovery and adaptation. Endurox, for example, has a 4-to-1 ratio—which happens to be the naturally occurring ratio in chocolate milk. In truth, "there's no magic ratio," says John Ivy, an exercise physiologist at the University of Texas at Austin and author of *Nutrient Timing: The Future of Sports Nutrition.* "Anywhere in the range from 2.5-to-1 to 4-to-1 works well," he says. For very long workouts lasting a few hours or more, a higher ratio of up to 6-to-1 may be appropriate, Stellingwerff adds.

Whatever the ratio, the idea that the body needs fuel immediately after exercise is now widely accepted. But some people take this advice too enthusiastically, as research at the University of Massachusetts at Amherst's Energy Metabolism Laboratory has shown. One of exercise's prime benefits for those seeking to lose weight is that it heightens insulin sensitivity, which helps to clear sugars from the bloodstream. Researchers at Amherst asked 16 sedentary, overweight subjects to walk on a treadmill for an hour a day at a moderate pace, burning 500 calories. Half the participants replaced the lost calories by drinking a sports drink and eating immediately after the workout, while the other half were given nothing. Surprisingly, while insulin sensitivity spiked 40 percent in the abstaining group, no improvement at all was seen in the group that refueled.

These results suggest that if you're trying to lose weight rather than trying to win races, you might be better off skipping

A POST-WORKOUT SNACK

How to get one gram of carbohydrate and 0.3 grams of protein per kilogram after exercise

- For a 120-pound person
  - A tuna sandwich and a 16.9-ounce sports drink
  - A cup of oatmeal with milk and an 8-ounce sports drink
- For a 175-pound person
  - A protein sports bar and a 25-ounce sports drink
  - Spaghetti with lean meat sauce and a cup of low-fat milk

the post-workout snack. That doesn't necessarily mean fasting completely: Ivy points to other studies showing that people who take in a small amount of protein after exercise are less likely to overeat several hours later. But the basic message is clear, Ivy says: "If you've gone out and burned 300 calories by walking for 30 minutes, don't refuel by taking a 500-calorie dietary supplement." Let the magnitude of your workout dictate the size of the chocolate milk carton.

How much should I drink to avoid dehydration during exercise?

It's the first lesson you learn about exercising in the heat: if you don't drink enough, you'll get dehydrated, and that will force you to slow down. By the time you feel thirsty, we're told, it's already too late. But some exercise physiologists, led by South African researcher Tim Noakes, have reached an alternative conclusion. It's not being dehydrated that actually slows you down, Noakes argues; rather, it's letting yourself get thirsty that signals to your brain to put the brakes on.

The idea that drinking according to thirst isn't enough to replenish your sweat losses is supported by plenty of research. For example, a 2007 study by researchers at the Gatorade Sports Science Institute found that experienced runners who were allowed to drink as much as they wanted during a 75-minute run replaced only about 30 percent of the fluid they lost by sweating. The conclusion the Gatorade scientists drew from this study is that we should plan to drink much more than thirst alone dictates—but Noakes disputes this interpretation.

The initial studies linking dehydration with impaired performance date back to World War II, as researchers sought to give U.S. military forces fighting in desert or jungle conditions an edge over their enemies. Since then, dozens more studies have shown that if you dehydrate someone (by administering a diuretic, for example), their exercise performance will suffer. Similarly, when subjects aren't allowed to drink during prolonged exercise, their performance is decreased. Based on these studies, researchers concluded that sweating out more than about 2 percent of your total body weight will slow you down.

But the subjects in all these studies are not just dehydrated; they're also forced to become thirsty. None of the studies show that subjects who drink freely according to thirst (and thus fail to replace their sweat losses) perform any worse than subjects who drink enough to replace all their sweat. Both thirst and the slowdown that eventually follows are the body's way of protecting itself in advance from damaging dehydration, Noakes argues. The flaw in the conventional studies is illustrated by a 2006 study in which subjects were told in advance whether or not their fluid intake would be restricted during a 50-mile cycling trial. When they knew they wouldn't be able to drink freely, the subjects biked more slowly right from the start of the trial, long before any physical effects of dehydration could have had an impact.

In Noakes's "central governor" theory (see p. 57), the brain monitors signals from various parts of the body with the goal of reducing exercise intensity before the body is in any danger of damaging itself. As fluid levels drop—but before they reach the point at which performance would be compromised—the central governor responds by initiating thirst and reducing intensity. In this picture, you won't slow down unless you actually feel thirsty, no matter how much fluid you've lost. Studies of endurance athletes show that the thirst mechanism varies greatly between individuals: some drink very little during races, while others drink a lot. Interestingly, the fastest finishers tend to also be the most dehydrated—a finding that lends support to Noakes's argument.

Until a few years ago, few researchers paid attention to Noakes's ideas. But the potentially fatal dangers of drinking too much—which were first pointed out by Noakes in the 1980s but ignored for nearly two decades—have caused a re-evaluation of hydration guidelines. Most authorities still recommend aiming to limit sweat losses to less than 2 percent of your body weight, but the consensus is no longer as strong. In that light, it's worth considering the advice Noakes gave in a 2007 article in *Medicine & Science in Sports & Exercise*: "Drink according to the dictates of thirst. If you are thirsty, drink; if not, do not. All the rest is detail."

Is it possible to hydrate too much?
In 2007, 28-year-old mother of three Jennifer Strange collapsed and died in her Sacramento home after taking part in a contest—"Hold your Wee for a Wii"—sponsored by a local radio

station. The goal was to drink as much water as possible without going to the bathroom, with the winner earning a Nintendo Wii. Strange's death was blamed on a condition called hyponatremia, sometimes known as water intoxication. Simply put, she had drunk so much that the levels of sodium in her blood were diluted, causing dangerous—and in this case fatal—swelling of the brain.

Over the last decade, medical directors at major marathons have sounded the alert about the dangers of overdrinking, thanks to about a dozen deaths at running races attributed to hyponatremia since the condition was first identified at the Comrades ultra-marathon in South Africa in 1981. Because the condition was so obscure, medical teams often mistook the symptoms for dehydration and pumped more fluids into collapsed runners, making it worse. Now that race personnel are aware of the dangers, these mistakes are less likely to occur.

Still, the condition is far more common than most people realize, since its first stages may not produce any obvious symptoms. Researchers at the University of London recruited 88 volunteers taking part in the 2006 London Marathon to give pre- and post-race blood samples in order to measure sodium levels. To their surprise, 11 of the volunteers (or 12.5 percent) developed asymptomatic hyponatremia, as indicated by abnormally low sodium levels—a particularly high number considering the cool and wet conditions of that year's race, which discouraged excess drinking. As expected, the runners who developed hyponatremia drank more frequently during the race (typically every mile) compared with those who stayed healthy (typically every two miles).

None of the runners in the study suffered any ill effects, but the high prevalence of low sodium levels suggests that many runners are still aiming to drink as much as possible along the route. Those who take more than four hours to complete a marathon are thought to be at higher risk, because they have more time to ingest water. Even though sports drinks have sodium in them, there's no evidence that they're less likely to cause hyponatremia. The best solution is simply to avoid drinking too much—no more than about eight ounces every 20 minutes, according to some experts. Or according to others (see p. 213), only when you're thirsty.

What ingredients do I really need in a sports drink?

If you're an old-school type who thinks plain water is all you need, consider this puzzling fact: rinsing your mouth with a drink containing carbohydrates will boost your athletic performance, even if you don't swallow and can't taste the carbs.

Of course, it's not just carbohydrates that you find in sports drinks these days. The latest offerings feature a bewildering array of formulations aimed at different sports and levels of activity, along with high-tech additives that purport to improve everything from alertness to metabolism. But you should be wary of the hype surrounding these magic ingredients. The core of any sports drink remains simple, says University of Guelph researcher Lawrence Spriet. Here are the three key ingredients, in order of importance:

- **FLUIDS:** The first point is simple: "If you're engaging in physical activity you're going to lose fluids," says Spriet, who also serves as chair of the Gatorade Sports Science Institute's Canadian advisory council. Although the precise link between dehydration and performance is still a topic of debate, it's clear that letting yourself

get thirsty during exercise will compromise your
performance.

- **CARBOHYDRATES:** The second element is carbohydrates, which are typically found in sports drinks in
  the form of glucose or other easily digested sugars. The
  goal is to maintain blood-sugar levels and replace glycogen stores in hard-working muscles, which is essential
  in bouts of exercise lasting longer than an hour. Sports
  drinks traditionally contain about 6 percent carbohydrate,
  about half the level of a typical juice or soft drink. That's
  about as high a concentration as your stomach can handle
  before absorption is slowed down, according to a recent
  study by University of Birmingham researcher Asker
  Jeukendrup in the journal *Nutrition & Metabolism.* Newer
  formulations such as Gatorade's G2 have cut the carb
  level to 3 percent, which is more appropriate for people
  who aren't pushing hard for hours at a time and thus
  don't need the extra calories.

  Scientists have long been puzzled about why carb-
  filled drinks also seem to help in shorter bouts of exercise,
  when energy stores shouldn't be an issue. Researchers
  at the University of Birmingham published a study in
  2009 showing that cyclists performed better in a time
  trial if they rinsed their mouth with a drink containing
  either glucose or a tasteless carbohydrate called maltodextrin but saw no improvement from rinsing with an
  artificially sweetened drink. Brain scans showed that
  glucose and maltodextrin made the reward centers in the
  subjects' brains light up while artificial sweetener didn't,
  suggesting that our mouths have previously unknown
  carbohydrate sensors. Sports scientists have now begun

advising elite athletes to rinse and spit out sports drinks as they approach the end of grueling endurance races, when their stomachs may be unable to handle drinking anything.

- **SALTS:** Electrolytes, which replace the salts lost in sweat, are thought by some researchers to play a role in muscle cramping (see p. 68) but are more relevant to post-exercise recovery for most people. "You have to be working really, really hard for the salt to matter," Spriet says. Gatorade makes a little-known product called GatorLytes, which is essentially a sachet of salts that you add to regular Gatorade to up its electrolyte content—but it's available only to high-level sports teams, since typical athletes simply don't need it.

For the average person working out at the gym for an hour or less at a time, there's no need to drink anything other than water. If you prefer a sports drink, choose one with a smaller amount of carbohydrate (3 percent is better than 6 percent), or simply dilute a standard sports drink with water. And don't put your faith in the magical claims of high-tech additives in some sports drinks—because beyond the three ingredients listed above the science gets a lot weaker. In fact, Gatorade's relaunch of its product line in the United States coincided with the decision to disband its U.S. scientific advisory panel in 2009. The new line boasts specialized formulations such as vitamin C to perk you up in the morning, B vitamins to help you metabolize energy, theanine to improve focus, antioxidants to "protect your body," and so on. Spriet is unimpressed. "Everyone wants to make things more complicated, but there's a reason the basic formulation hasn't changed in years," he says. "It's fluid, sugar, and salt. That's all it is—and it works!"

Will taking antioxidant vitamins block the health benefits of exercise?

Every year when cold and flu season hits, sales of orange juice soar as people seek the protection of vitamin C. Faith in the power of antioxidants is deeply entrenched. But over the past few years, a series of vast studies involving hundreds of thousands of subjects has failed to find any health benefits from antioxidant supplements. Now, another group of studies suggests that popping these pills may even block some of the benefits of exercise and slow down post-workout muscle recovery. It would be premature to pronounce the end of the vitamin era on the basis of a few studies—just as premature as it was to leap on the vitamin bandwagon in the first place. But some skepticism is due. "For something like vitamin C, it's important to have enough," says Stephen Cheung, a physiologist at Brock University in St. Catharines, Ontario. "But that doesn't mean more is better."

Antioxidants—vitamins C and E plus molecules ranging from beta carotene to currently fashionable resveratrol—attack and neutralize the "free radicals" associated with aging and disease. Exercise stimulates the production of free radicals, which is why athletes are often advised to take extra antioxidant supplements. But exercise itself is also an antioxidant. During exercise, the body gradually learns to produce more and more of its own antioxidants in response to the spike of free radicals generated by working out. One theory now gaining support is that taking extra antioxidants means that the body never gets the opportunity to adapt on its own.

In 2009, Michael Ristow and his colleagues at the University of Jena, in Germany, published a study in the *Proceedings of the National Academy of Sciences* examining how a four-week exercise program affected insulin sensitivity—one of the most significant health benefits conferred by physical activity. Half of the 40 volunteers were given a placebo and saw significant improvements in insulin sensitivity; the other half took 1,000 mg of vitamin C and 400 IU of vitamin E each day and saw no change despite the exercise regime. To Ristow, this suggests that antioxidants are unequivocally bad, even though the research in favor of eating fruit and vegetables is unimpeachable. "This insinuates that fruit and vegetables are healthy despite their content in antioxidants," he explains, "[so] other compounds in fruit and vegetables are responsible for their health-promoting effects."

The idea that antioxidants can stave off some of the muscle damage and soreness caused by free radicals after heavy exercise has also taken a hit. In 2009, researchers studying the Portuguese national kayak team found hints that, compared with a placebo, a cocktail of antioxidants actually delayed muscle recovery after training. Victor Hugo Teixeira of the University of Porto, the study's lead author, speculates that free radicals may serve as a natural brake to stop you from pushing too hard. Taking antioxidant pills could override that brake, allowing your muscles to work a little harder and sustain greater damage. If that's true, athletes might gain an edge from taking antioxidants right before a competition but would suffer from impaired recovery if they took them on a regular basis.

Even if antioxidants did ruin your workout, many people would gladly take that risk if it helped them avoid the flu. It's well established that antioxidants can help boost immune function in people who have undergone truly extreme physical exertion, like running an ultra-marathon, Cheung says. But it's less clear that the same benefits accrue in everyday life. In a study published last year, Cheung had volunteers cycle at moderate intensity for two hours—hardly slacking—and tested whether their immune function was helped by 1,500 mg of vitamin C a day for two weeks afterwards. The results were equivocal: if there was any effect, it was weak.

Cheung's advice is to ensure you're getting enough vitamin C from your diet—and if not, to change your diet before resorting to supplementation. In a field where the science is still hotly contested, this seems like wise counsel. Someday, perhaps, we'll know exactly which molecules make fruit and vegetables so good for us—but until then, as long as you're eating lots of them, you don't have to worry about which ones.

Should I be taking probiotics?

Over the last few years, grocery store shelves have been taken over by "helpful" bacteria. Particularly in the dairy section, foods now trumpet the presence of live cultures and the health benefits they offer. The general term *probiotics* refers to live micro-organisms that interact with the existing bacteria in your gut to produce a positive effect on your health. Some of the most common examples are bacteria that feed on lactose and are used in the fermentation process to produce yogurts and cheeses.

A number of studies have shown that certain probiotic strains can help boost immune function. For instance, antibiotics often kill off some of the beneficial bacteria that live in your gut, an outcome that can leave you susceptible to gastrointestinal problems like diarrhea. A recent analysis of 34 different studies concluded that several strains of probiotics (*S. boulardii,*

L. rhamnosus, L. acidophilus, and *L. bulgaricus*) all reduced the risk and severity of antibiotic-related and traveler's diarrhea. There's also some evidence that probiotics can enhance the mucous lining of the respiratory tract, to help prevent viral infections like colds and coughs.

Because the extreme effort associated with something like marathon training can leave your immune system temporarily depleted (see p. 23), athletes have been particularly interested in the potential benefits of probiotics. In a 2008 study at the Australian Institute of Sport, 20 elite runners spent four months of winter training taking capsules containing either *Lactobacillus fermentum* or a placebo. By the end of the study, the number of days during which the runners reported symptoms of respiratory infection was 2.4 times higher in the placebo group than in the probiotic group, and the symptoms were more severe on average in the placebo group. The researchers also took blood samples that showed elevated levels of interferon gamma, a marker of immune function, in runners taking the probiotic.

Another study monitored 141 runners training for the Helsinki marathon, giving them capsules containing either *Lactobacillus rhamnosus* or a placebo for three months leading up to the race and monitoring them for a further two weeks afterwards. In this case, there was no difference in the number of respiratory infections or gastrointestinal "episodes" reported by the two groups. However, there was a trend for GI problems to clear up more quickly in the probiotic group (2.9 days for problems before the marathon and 1.0 day after the marathon) compared with the placebo group (4.3 days before and 2.3 days after).

These results are definitely encouraging—the problem is that every individual strain of probiotic bacteria has different effects, and there's not yet any consensus about which ones are best or how much we need to take. Fortunately, this is one of those cases where it makes sense to incorporate foods like yogurt with live bacterial cultures into your diet even though the scientific research is still incomplete. Even if the probiotics don't do anything for you, you'll still have eaten a bunch of nutritious (and tasty) food.

Will vitamin D make me a better athlete?

In a 2009 study, researchers from the University of Manchester in Britain asked 99 adolescent schoolgirls to perform a series of one- and two-legged jumps, then took blood tests to see how much vitamin D they had in their bodies. There was a clear correlation: the more vitamin D, the higher, faster, and more powerful the jumps. To many, this was confirmation of what they'd suspected for some time: the "sunshine vitamin" could turn out to be the ultimate natural performance enhancer. But it's not quite that simple.

D has been the star vitamin of the past few years, piling up study after favorable study even as the claims of its fellow vitamins are steadily being debunked. According to various studies, vitamin D fights cancer, builds bones, combats heart disease, tunes up your immune system, and provides a long list of other benefits. Since it's produced in the body as a response to ultraviolet light from the sun, people who live far from the equator are particularly at risk of deficiency in winter—which may explain why diseases like lung cancer and breast cancer are most likely to kill you if you're diagnosed during those gloomy months. You can get some vitamin D from sources like fatty fish and fortified milk, but the vast majority comes from either sunlight or supplements.

Interest in the sun's potential as a performance booster dates back at least to a rudimentary Russian study in 1938 in which four students improved their 100-meter dash time by 7.4 percent after a course of UV radiation, while controls improved by only 1.7 percent. In subsequent decades, German researchers also tried boosting performance using sun lamps and identified vitamin D as the probable cause. But this research petered out in the 1960s without any rigorous conclusions. Other studies have looked at vitamin D's links with parameters like reaction time and muscle protein synthesis. But, according to a 2009 review of the topic in *Medicine & Science in Sports & Exercise,* no studies have ever looked for direct links between athletic performance and levels of vitamin D indicated by blood tests.

Much of the debate centers on how you define *deficiency.* According to a 2008 study in the *American Journal of Clinical Nutrition,* vitamin D levels in American children and adults appear to have declined since the 1980s, possibly because people spend less time in the sun and drink less milk. About half of adults now have sub-optimal levels of vitamin D, according to the study. Notably, three-quarters of the girls in the University of Manchester study were found to be vitamin D deficient, which makes it less surprising that higher levels improved jumping performance. After all, even a glass of water is performance-enhancing if you're thirsty.

Several large-scale studies involving thousands of people are now in progress to untangle the cause-and-effect links between vitamin D and various diseases. Despite the uncertainty, there's enough evidence to suggest that you should be aware of your vitamin D levels and make sure you either take supplements or get enough sun. Once you reach "normal" levels, there's currently no evidence that further vitamin D will make you a better athlete—but going from deficient to normal could definitely put a spring in your step.

Is there any benefit to deliberately training with low energy stores?

One of the hottest controversies in current sports nutrition was sparked by an unusual Danish study published in 2005. Volunteers performed a 10-week training program in which they exercised one leg every day and exercised the other leg twice as much every second day. That meant that the leg trained every other day did half of its workouts in a highly fatigued state, having been depleted of glycogen by the first half of the workout. By the end of the study, this leg had developed significantly greater endurance, giving rise to a new concept that was soon dubbed "train low, compete high," in which athletes seek to do part of their training when their energy stores are greatly depleted ("training low") so that they'll perform even better when they're fully fueled ("competing high").

There's no doubt that having full carbohydrate stores improves your endurance (see p. 206). In fact, that's the point: "training low" is the nutritional equivalent of wearing a weighted vest to make your workout harder. There have long been rumors that athletes like Miguel Indurain, the five-time Tour de France champion, tried this approach by doing some of his training in a fasted state. But there's been little evidence that it actually works: even the Danish study had several flaws—notably that the subjects were untrained, which makes it much easier to observe improvements in performance, and that "single-leg kicking" isn't an activity that has any particular relevance to the real world.

Several recent studies have tried similar protocols with trained cyclists, and they've found that training low really does stimulate the body to adapt differently—but it doesn't seem to produce any actual performance benefits. For example, a 2010 study at the University of Birmingham used highly invasive muscle biopsies and isotope tracers to measure the different muscular and metabolic changes produced by training high and training low. As expected, they found that training low taught the body to burn more fat instead of carbohydrate, which should theoretically improve endurance performance by allowing carbohydrate stores to last longer before running out. But in a one-hour time trial, there was no difference between the two groups.

This apparent contradiction is similar to discussions of the "fat-burning zone" for weight loss (see p. 190). In both cases, researchers have figured out how to make the body rely more on fat instead of carbohydrate—but you don't lose more weight or bike faster, because the body seems to compensate for the change. In fact, there's some evidence that in increasing your fat-burning abilities, you also harm your carbohydrate-burning capacity. It's tempting to believe that this doesn't matter in ultra-endurance events like marathons or 100-mile bike races, where fat-burning plays a major role. "However," Australian Institute of Sport nutritionist Louise Burke pointed out in a 2007 commentary, "the strategic activities that occur in such sports— the breakaway, the surge during an uphill stage, or the sprint to the finish line—are all dependent on an athlete's ability to work at high intensities that are carbohydrate-dependent."

In practice, there are two approaches to training in a carbohydrate-depleted state. One is the approach used in the studies described above: deplete your muscle glycogen stores with 30 to 60 minutes of moderate exercise at about 70 percent of maximum effort. Then, without refueling, do some harder training. Alternatively, you can try working out first thing in the morning without eating anything first, possibly having eaten a low-carbohydrate dinner the night before, so that your whole body is low on glycogen. Both these strategies can be very stressful for the body and shouldn't be attempted more than once or twice a week. Recovery afterwards is crucial, including lots of carbohydrates.

At this stage, although "train low, compete high" has become a popular buzz-phrase, the research remains highly uncertain. For most people, the best bet is to let others do the risky and often unpleasant experimentation—then, if it does turn out to provide measurable performance benefits, give it a try once the details have been worked out.

Can I get the nutrients I need for a heavy exercise regimen from a vegetarian or vegan diet?

Running 167.7 miles in a single day, as 36-year-old Scott Jurek did in setting a new American record for the 24-hour run in 2010, is a staggering feat by any definition. But Jurek's accomplishment garnered extra attention because he follows a strict vegan diet. How was it possible, people wondered, to run over 140 miles week after week and consume 5,000 to 8,000 calories a day, with no meat or animal products?

For decades, would-be vegetarian athletes have been warned about the shortcomings in their diets, like the possible shortage of essential elements such as protein, iron, and calories. But only a few isolated studies have compared the actual performance of vegetarian and omnivorous athletes, with generally favorable results. One in 1970 found no difference in lung function

HIGH-PROTEIN PLANTS

Spinach (3 cups, cooked): 15 g of protein
Asparagus (3 cups, cooked): 12 g
Lentils (1 cup, cooked): 18 g
Oats (½ cup, dry): 13 g
Quinoa (1 cup, cooked): 8 g

and thigh muscle size between the two groups. A 1986 Israeli study found no difference in serum protein between vegetarian female athletes and matched controls, and a 1989 German study found no difference between vegetarians and non-vegetarians in the finishing time of a 1,000-kilometer run.

It is, of course, entirely possible to consume an unhealthy and deficient vegetarian diet. If you took a typical North American diet and simply removed the meat from it, you'd almost certainly not get enough protein to support a heavy exercise regimen. But if you take the time to consume good sources of vegetable protein, you'll have little trouble meeting your protein needs (see p. 117). Similarly, getting enough calories is simply a matter of eating more. "The first thing to worry about isn't so much what you eat, but how much you eat," Jurek explained to a *New York Times* reporter. "You have to take the time to sit at the table and make sure your calorie count is high enough."

There are, however, some special considerations that vegetarians and vegans need to be aware of. Though leafy greens like spinach and kale are excellent sources of iron, only about 10 percent of iron from plant sources can actually be absorbed by the body (compared to 18 percent from meat). Female endurance athletes, in particular, are prone to low iron levels, so they may need to consider iron supplements if tests show their levels are low. In addition, a 2010 review in the journal *Current Sports Medicine Reports* by Joel Fuhrman and Deana Ferreri identified several other micronutrients that vegan and vegetarian athletes may be deficient in. In particular, zinc, vitamin B12, and the omega-3 fatty acid DHA are all crucial for physical performance and are either hard to absorb or hard to get enough of from plant sources, so they recommend taking supplements.

"Clearly," Fuhrman and Ferreri conclude, "a properly designed vegan (or near-vegan) diet can meet the nutritional demands of a speed and agility athlete, such as tennis, skiing, basketball, track, and soccer, but may not be ideal to maximize growth over 300 lb as a football linebacker." Scott Jurek might disagree with this conclusion, since he manages to consume up to 8,000 calories a day—but there's no doubt that such extreme consumption requires a special and perhaps rare dedication. Most of us, though, have no desire to exceed 300 pounds or run for 24 hours at a time, so a well-balanced vegan or vegetarian diet is perfectly capable of meeting our needs.

CHEAT SHEET: NUTRITION AND HYDRATION

- To carbo-load, you can max out your carbohydrate stores with just one day of eating 10 grams of carbohydrate per kilogram of body weight.
- To avoid stomach problems during exercise, avoid high-fiber and high-fat foods in pre-exercise meals and allow at least three hours for digestion.
- Eat as soon as possible after exercise (within at most two hours) to enhance recovery and training gains, aiming for a roughly 4:1 ratio of carbohydrate to protein.
- Losing more than 2 percent of your body weight in fluid is thought to hurt performance, but some scientists now believe that simply drinking when you're thirsty is sufficient.
- Drinking too much can lead to dangerously low sodium levels (hyponatremia) and possibly death. Don't drink more than eight ounces every 20 minutes.
- For events lasting longer than an hour or two, consume fluids containing no more than 6 percent carbohydrate and some electrolytes.
- Antioxidants like vitamins C and E may block some of the health effects of exercise and slow post-exercise muscle recovery, though the evidence is still preliminary.
- Probiotics can help ward off respiratory infections and digestive problems, but it's not yet clear which strains of bacteria are best and how much is required.
- Vitamin D (the sunshine vitamin) is crucial for good health and many people require supplementation, but there's no evidence that extra vitamin D improves athletic performance if you're not deficient.
- Training with depleted carbohydrate stores (e.g., before breakfast) can teach your body to burn more fat and store more carbohydrates, but there's little evidence so far that it boosts performance.
- Studies have found no difference in physiology or results in athletes with vegetarian and non-vegetarian diets, provided that the diets are balanced and include all necessary nutrients.

11: Mind and Body

A RESEARCHER IN WALES IS CURRENTLY trying to improve the stationary bike performance of a group of volunteers using a program of vigorous training, 45 to 90 minutes a day, five days a week . . . playing a cognitively challenging computer game. Training the brain, he believes, will translate into improved physical performance. However the results turn out, the fact that such an experiment isn't simply laughed out of the lab shows how much our understanding of the link between brain and body has changed in the past decade.

For decades, exercise physiology has struggled to pin down the limits of physical performance by studying the heart, lungs, and muscles of athletes. New experiments show that the brain plays a fundamental and often surprising role. And it also works the other way around: exercise shapes your brain, stimulates growth, and enhances memory and cognition—and some types of exercise are better than others.

If my brain is tired, will my body's performance suffer?
Many people spend their workdays sitting on a chair in front of a computer, with physical exertion limited to an occasional stroll down the hallway. From a purely physical perspective, it's hard to imagine a cushier existence. But if you head to the gym after work, you might notice that your performance suffers

when you've spent the afternoon wrestling with a particularly complicated problem. In 2009, researchers from Bangor University in Wales published the first rigorous investigation of this phenomenon in the *Journal of Applied Physiology*. They asked 16 volunteers to ride a stationary bicycle to exhaustion under two different conditions: once after taking a challenging 90-minute cognitive test and once after passively watching 90 minutes of documentaries about trains and cars. Sure enough, the subjects who had been using their brains reached exhaustion after 10 minutes and 40 seconds on average—a full minute and 54 seconds before the movie-watchers.

It's possible that thinking actually burns up enough energy to affect your physical performance: the test-takers recorded slightly higher heart rates (65 beats per minute) compared with the movie watchers (62 beats per minute), likely due to the brain's heightened demand for glucose. But the data taken during the cycling test suggest that this wasn't a significant factor. The two groups produced identical physiological responses to exercise, as measured by quantities such as heart rate, oxygen consumption, and blood lactate. The only difference was in their subjective rating of how hard the exercise felt: right from the moment they started pedaling, the mental-fatigue group rated their perceived exertion a few points higher on a scale of 6 to 20. Both groups stopped pedaling when their perceived exertion hit 20—which means the mental-fatigue group stopped earlier, with lower physiological measures of fatigue than the control group.

For the researchers, this finding offers powerful evidence that our traditional understanding of exhaustion as the result of reaching some physical limit in the body's capabilities is flawed. "Overall, it seems that exercise performance is ultimately limited by perception of effort rather than cardiorespiratory and musculoenergetic factors," they wrote. Intriguingly, our perception of effort during exercise is regulated in a part of the brain called the anterior cingulate cortex—which is precisely the part of the brain activated by the 90-minute cognitive test used in the experiment. This suggests that a bout of hard thinking might disrupt our sensation of physical effort, even though the muscles themselves are completely unaffected. The most eyebrow-raising implication of this study is that it might be possible to improve physical performance through brain training, without leaving your armchair, by postponing fatigue of the anterior cingulate cortex. The Bangor research team is in the process of testing this remarkable hypothesis.

For elite athletes, the study highlights the benefits of a period of mental relaxation before particularly hard workouts or competitions. In general, though, the researchers are careful to point out that exercising after work is still a great idea: the effects of mental fatigue will have only a minor effect on moderate exercise, and the workout is a great source of stress relief. Still, it's worth bearing in mind that if you've been pulling an Einstein at work, you might need to cut yourself some slack at the gym.

Does it matter what I'm thinking about when I train?

For many people, heading out for a run or a bike ride offers a mental break—a chance to think about the events of the day, or about nothing at all, while the legs navigate on autopilot. But for those who are looking to lower their best race times, a growing body of research suggests that what's going on in your head during training sessions can make a big difference in how effective those sessions are. "It's not just physical intensity that counts, it's mental intensity," says Joe Baker, a researcher at York University in Toronto.

Over the past few decades, psychologists have reached the remarkable conclusion that your level of achievement in fields ranging from sports to music to science depends less on natural talent than on the number of hours you spend doing "deliberate practice," a term coined by Florida State University cognitive psychologist Anders Ericsson. In one of his seminal studies, Ericsson found that the virtuosos at major philharmonics had averaged 7,400 hours of deliberate practice by the age of 18; typical professionals had averaged 5,300 hours; and those who ended up teaching violin instead of performing had spent only 3,400 hours.

Not all practice is "deliberate" practice. Rather than simply repeating tasks over and over, it involves setting specific goals and monitoring how well you perform, constantly adjusting and improving your technique. This seems like the opposite of most training for endurance sports—heading out the door and running for an hour at a comfortable pace, say, with no specific goals, minimal feedback, and no thought about technique.

But top endurance athletes rely on a number of training techniques that do fit the definition of deliberate practice. In a study of the training practices of elite runners by University of Ottawa researchers Bradley Young and John Salmela, what separated the highest-performing group from their less accomplished peers was how much they incorporated elements like interval training, tempo runs, and time trials, all of which require ongoing attention to pace and other feedback. "High quality and high intensity, rather than long slow distance, is at the heart of deliberate practice," Baker says.

Deliberate practice also involves monitoring your training and progress over the course of weeks and months and making appropriate adjustments, rather than just doing what you've always done. A study of Ironman triathletes by Baker and his colleagues found that the most experienced athletes planned their training year carefully, taking regular easy weeks to allow their bodies to recover so that they could steadily build their training to a peak. The novices, in contrast, simply trained as hard as they could until cumulative fatigue or injuries forced them to back off, resulting in less training overall.

Traditionally, researchers have divided the mental strategies used by endurance athletes into "associative" and "dissociative." When you're associating, you're concentrating on the task at hand: your breathing, your pace, and so on. When you're dissociating, you're thinking about anything but the task at hand: the weather, or last night's TV show. A series of studies over the past few decades has demonstrated that faster runners have more associative thoughts during competition than their slower rivals, who have more dissociative thoughts. "But there's an important message," Baker notes. "No one has suggested that top runners associate all the time."

Similarly, psychologists don't suggest that you try to make all your training deliberate. Even the virtuoso violinists, famed for spending 10 or more hours a day practicing, managed to average only a few hours a day of deliberate practice. For most people, the majority of exercise time should remain relaxed, a mental diversion. But adding a segment of deliberate practice a few times a week could make a big difference in your race performance. And there may be an added bonus. Young and Salmela's study of elite runners produced one very unexpected result: they found that the types of training that took the most effort and concentration—the most deliberate, in other words—were rated as the most enjoyable sessions by the runners. So deliberate practice may be hard, but it's satisfying—especially on game day.

Does listening to music or watching TV help or hurt my workout?

In a study published in 2009, researchers at Britain's Liverpool John Moores University secretly sped up or slowed down music by 10 percent and observed the effect on subjects riding exercise bikes. Sure enough, like marionettes on musical strings, the riders unconsciously sped up or slowed down. The study confirms what gym-goers have known for a long time: music can have a significant impact on your exercise performance, and its influence extends far beyond the simple psych-up provided by motivational lyrics.

The dominant theory about why music boosts exercise performance is that you have a limited ability to pay attention to the information your senses gather. Focus on sounds and sights, the theory goes, and you're less aware of the distress signals your muscles are sending you. A 2007 study by Vincent Nethery of Central Washington University offered support for this theory. Subjects exercising at a constant workload reported less discomfort when listening to music or watching a video. In contrast, an earlier study by Nethery found that subjects wearing earplugs and a blindfold reported greater levels of discomfort during exercise, presumably because they had nothing to focus on except their fatigue.

The British study adds a new twist by controlling the factors that usually confound studies of music and exercise: personal preferences, volume, pitch, duration, genre, lyrics, and so on. The researchers chose six tracks that "reflected current popular taste among the undergraduate population" and combined them into a single 25-minute program, then digitally altered it to create faster and slower versions without changing the pitch. The subjects exercised to the three versions with a week in between each session, and none of them noticed the differences in tempo. A 10 percent difference is quite small, lead researcher Jim Waterhouse says: "Compare the interpretations of Beethoven symphonies by Toscanini and Klemperer, for example."

The link between musical tempo and effort confirms the findings of several earlier studies, albeit with greater rigor. What's new, though, is the fact that the subjects reported greater enjoyment and higher levels of perceived exertion after the sped-up session. In other words, the faster music didn't simply distract them from their discomfort; it motivated them to happily endure greater levels of discomfort.

This result—along with many other apparently conflicting studies in this area—suggests that the music-as-distraction theory ignores broader psychosocial factors, says Costas Karageorghis of Brunel University in Britain. He identifies rhythm, "musicality," cultural impact, and the external associations of the song as the four key factors influencing a listener's response. As a result, attempts to find universal effects of particular pieces, styles, or even speeds of music on exercise are doomed. Instead, Karageorghis says, people should tailor their play lists to their personal preferences, and gyms should play different types of music by the cardio machines (faster) and by the weights (motivational lyrics).

Interestingly, some researchers have found preliminary evidence that watching TV or videos is likely to slow your workout down, suggesting that too much distraction not only dulls your pain but also distracts you from putting forth an honest effort. The difference between video and music may have something to do with the active attention required to watch a video—holding your head in the right position, for example—compared to

HEADPHONE DANGERS

In 2007, headphones were banned at all major road races in the United States. Although that ban has since been relaxed, listening to music while running or biking remains a serious safety concern on roads and paths with traffic—be it cars, bikes, or simply other pedestrians. On empty trails where no one else is around, it's still crucial to be aware of your surroundings, particularly for women exercising alone. The performance-boosting effects of music are best enjoyed indoors.

passively listening to music, Nethery says. "It may also highlight the value of rhythm associated with music," he adds. That would mean listening to a podcast or to talk radio would be more akin to watching a TV show than listening to music, though this hypothesis has yet to be tested.

Of course, even the "wrong" distractions are still beneficial if they get you out the door or keep you exercising longer. But it's worth being aware of the effects. Try paying attention to how different songs make you feel and perform during different exercises, and you may learn how to give yourself a boost when you most need it.

Will I perform better under pressure if I focus harder?

When you're lining up a crucial putt, the last thing you want to hear is an impatient jerk in the foursome behind you yelling "Hey buddy, could you hurry it up a bit?" But new research suggests that the jerk might be doing you a favor. Psychologists and neuroscientists are finding that, when you perform complex motor sequences that you're very familiar with, concentrating too much on the details of the task makes your performance worse. Too much concentration is what causes choking on the putting green or at the free-throw line—and it's why a bit of distraction can be a good thing.

Even skills as simple as tying your shoes begin as a sequence of separate actions that you consciously execute in a certain order, explains Clare MacMahon, a Canadian researcher now at Victoria University in Australia. As you master the skill, the steps become combined into a single action that is stored in your "procedural" memory, beyond conscious control. "If you then try to consciously monitor yourself—'Make two bunny ears, one bunny goes around the tree' and so on—you break the skill back down into its components," MacMahon says, "and you end up looking like a novice again."

MacMahon was a co-author of a 2002 study in the *Journal of Experimental Psychology* that compared novice and expert soccer players dribbling a ball while either focusing on their technique or being distracted by sounds. Novices performed better when they focused on technique, but experts—for whom technique had progressed to subconscious control—performed better when they were distracted. As an added twist, the researchers repeated the experiment with the subjects using their non-dominant foot. In this case, both novices and experts performed better while focusing on technique: even the best players weren't truly expert with their non-dominant feet.

Sian Beilock, the University of Chicago researcher who was the lead author of the soccer study, published another study in 2009 in which novice and expert golfers were asked to putt either as quickly as possible while maintaining accuracy or taking as much time as they wanted. Again, novices and experts responded differently. The novices performed best when they took their time, while the experts performed best when they were told to hurry up. "It's really clear that paying too much attention to your step-by-step actions can disrupt fluidness," Beilock says. "And this is what happens under stress, what people call 'choking.'"

Other researchers have extended these results to basketball, hockey, and even darts. More surprisingly, a 2009 study in the *Journal of Sports Sciences* found that focusing too intently was even disruptive in running, a task whose mechanics most people consider fairly straightforward. Researchers found that runners focusing on their form or their breathing consumed more oxygen and burned more energy than runners who simply watched their surroundings. If nothing else, this is a reminder that walking and running are complex tasks that your subconscious mind executes very efficiently. ("If you don't believe walking is complex, why can't we program a robot to walk normally?" Beilock points out.)

Of course, you still need your conscious mind to master skills in the first place. But once you've become proficient, it's worth remembering that, when the pressure is on, scrunching up your forehead and focusing on your mechanics is more likely to hurt you than help you. And you can also use this knowledge as an offensive weapon, MacMahon points out. If you're playing tennis and your opponent aces you, compliment them: "Wow, that was a great serve. Are you doing something with your wrist?"

Can swearing help me push harder in a workout?

If you're looking for that extra edge that will allow you to lift one more rep or maintain your pace near the end of a tough workout, consider the latest research from psychologist Richard Stephens of Keele University in Britain. After hitting his thumb with a hammer, Stephens let loose with a string of expletives—a common enough occurrence, but one that left him wondering why humans have this nearly universal habit of "cathartic swearing." To find out, he asked 67 volunteers to dunk their hands in ice-cold water and keep them there for as long as possible. Half of them were told to yell a word from their list of "five words you might use after hitting yourself on the thumb with a hammer," while the other half chose a word from their list of "five words to describe a table." Sure enough, swearing significantly increased the length of time subjects could withstand the pain, by 30 percent for men and 44 percent for women—a difference that may have something to do with the fact that women swear less often, Stephens speculates. Swearing also raised heart rates and decreased perceived pain, again with a greater effect in women than men.

The results were actually the opposite of what Stephens expected. Pain theorists had believed that swearing was a form of "catastrophizing" or exaggerating the severity of the pain, but Stephens's results, which appeared in 2009 in the journal *NeuroReport*, suggest that something else is happening. Instead, it may be that swearing triggers feelings of aggression that allow us to tap into our fight-or-flight mechanism, pumping adrenaline through our veins and blocking pain. Similarly, Stephens points out, sports coaches often psych their players up with pre-game speeches laden with profanity. Whatever the precise mechanism, the finding that swearing increases pain tolerance explains why evolution has ensured that the behavior survives in virtually all cultures. Most language is generated in the left brain, but swearing appears to arise in the older emotion centers of the right brain: the limbic system and the basal ganglia. So the next time you're trying to get through the painful part of a race or workout, remember that you have the code words to access this primitive part of your brain—as long as there aren't any children within earshot.

(Strangely, this isn't the only study to suggest that expressing your inner jerk can boost your physical performance. In 2010, Harvard University psychologists reported that doing a good deed like giving money to charity, or even imagining doing a good deed, enabled volunteers to hold up a five-pound weight for longer than they could when thinking neutral thoughts. But they gained even greater strength from imagining themselves doing evil deeds like harming someone else—even without swearing!)

Is there such a thing as "runner's high"?

For decades, the "runner's high" was the Yeti of exercise science: an exciting phenomenon that lots of people claimed to have seen but that couldn't be reproduced on demand and couldn't be reliably documented or measured. The lucky ones described feelings of exaltation, effortlessness, and intense euphoria, usually during or after long or difficult runs, and some attributed it to the mood-altering effects of brain chemicals called endorphins. But skeptics weren't convinced: University of Michigan researcher Huda Akil, the president of the Society for Neuroscience, dismissed the idea as "a total fantasy in the pop culture" in a 2002 interview with the *New York Times*.

Endorphins are endogenous morphines, chemicals produced naturally within the body that alter mood and block pain much like opiate drugs such as morphine. Researchers have known for decades that vigorous exercise raises endorphin levels in the blood, but the brain is almost completely isolated from the blood in the rest of your body by a "blood-brain barrier." It was only in 2008 that researchers at the University of Munich, using new techniques, were able to perform a direct test of endorphin levels in the brains of runners who had just completed a two-hour run. Sure enough, endorphins were released in the brains of the exhausted runners—and more importantly, there was a direct correlation between the level of endorphin activity and the level of euphoria reported by the runners (who didn't know what the study was trying to measure). The areas of the brain most affected were the limbic and prefrontal areas, which are associated with mood.

Despite its fame, runner's high is actually quite rare, at least in its most extreme euphoric form. But some researchers believe that post-exercise endorphins have subtler effects that are far more widespread than previously realized—and that the opioid rush explains why some people become effectively addicted to exercise, persisting in their daily ritual even when they're injured or unhealthy. In support of this theory, psychologists at Tufts University administered naxolone, which blocks the action of opioid drugs, to a group of rats who were accustomed to vigorous exercise. The rats displayed symptoms of withdrawal, including teeth chattering and "wet dog shakes"— exactly as morphine-addicted rats would have under the same circumstances.

Of course, most people manage to find a happy medium: they may not experience the euphoric thrills of runner's high, but neither are they desperately jonesing for their next hit of exercise. Instead, the endorphins produced during their workout contribute to the general sense of calm and well-being that typically follows exercise—and that's enough to keep them coming back for more.

Will taking a fitness class or joining a team change my brain chemistry during workouts?

Consider the similarities between a modern exercise class and an ancient religious rite—the wise leader guiding the group through a series of ritualized movements, in perfect synchronization. If you're struggling to keep faith with your fitness goals, this apparent coincidence might offer a solution. New research suggests that group exercise unleashes a flood of chemicals in the brain, triggering the same responses that have made collective activities from dancing and laughter to religion itself such enduring aspects of human culture. For some (but not all) people, finding workout buddies could help turn fitness into a pleasant addiction.

In a 2010 issue of *Biology Letters*, researchers from Oxford's Institute of Cognitive and Evolutionary Anthropology reported on a study of the university's famed rowing team. The crew was divided into teams of six, each of which performed a series of identical workouts on rowing machines. The only variable was whether the workouts were performed alone or in teams with the six rowing machines synchronized by the crew's coxswain. After each workout, a blood-pressure cuff was tightened around one arm of each subject until he reported pain, an indirect method of measuring endorphin levels in the brain. Endorphins—the same chemicals that stimulate runner's high (see p. 243)—produce a mild opiate high and create a sense of well-being as well as blocking pain. Sure enough, the rowers' pain threshold was consistently twice as high after exercising with their team-mates compared with exercising alone, even though the intensity of the workouts was identical.

So where does this magic come from? The endorphin surges can likely be traced back to the evolutionary benefits of group bonding, the researchers suggest. Earlier studies have indicated that synchronized physical activity elevates mood and is associated with greater altruism. But synchronization is probably not the only factor involved, notes lead author Emma Cohen, who is now at the Max Planck Institute for Evolutionary Anthropology in Germany. "We also suspect that shared goals—ultimate goals, like winning the big race, and proximate goals, like endeavoring to row together in synch—are at least part of the trigger," she explains. Cohen is following up on this question by studying religious drummers in Brazil, while her former colleagues continue to experiment with the Oxford rowing crew.

Endorphins are produced by virtually any vigorous physical activity, as confirmed by the fact that even the solo rowing sessions in Cohen's study enhanced pain threshold to some degree. But group work appears to enhance the effect dramatically—and there's plenty of evidence that exercise classes meet that description. In a series of studies stretching back more than a decade, University of Saskatchewan professor Kevin Spink has found those who feel a greater sense of "groupness" and cohesion within an exercise class are more punctual, have better attendance, and even work harder.

Of course, not all collections of individuals qualify as a group. Spink and other researchers have identified factors that make some crowds "groupier" than others, such as the existence of group norms. For example, the shift in the past decade from sign-up exercise classes to drop-in classes has made it more difficult to build cohesion in these groups. Still, it appears that the most important factor is what's in your head, even for drop-in classes. "As long as I perceive the people I'm exercising with as a group, my adherence is way better," Spink says.

There is an important caveat regarding individual preferences. About a third of people enjoy exercising in groups; another third prefer exercising alone, while the remaining third
are indifferent, Spink notes. For those who are happy exercising alone, there's no reason to join a group. For everyone else,
exercising with partners or in groups has all sorts of benefits
that have nothing to do with neuroscience, from the simple act
of committing to meet someone to the pleasures of gossiping
during a workout. But the endorphin findings help explain
how exercise is transformed from a chore to a lifelong habit,
and indeed a pleasure, for some people—and suggest one way
of getting there.

What are the effects of exercise on the brain?

The theme of much of the research described in this chapter
is how much influence your brain has on the way you exercise. But it works the other way around too: the exercise you
do has wide-ranging effects on your brain, with the power to
alter mood, memory, and even the structure of the brain itself.
Over the long term, there's not much doubt that exercise makes
you smarter. Studies in rodents have shown that physical activity makes brains develop denser and more complex connec-
tions between neurons and stimulates the growth of new brain
cells. These effects are especially important during adolescence
and early adulthood, when your central nervous system is developing rapidly and taking the shape it will maintain, more or
less, for the rest of your life.

A massive Swedish study published in 2009 combed
through the records of 1.2 million 18-year-olds who had taken
compulsory military screening exams between 1950 and 1976.
The first finding was that aerobic fitness, but not muscular

strength, was associated with greater intelligence. But it wasn't just *being* fit that helped—*getting* fit also offered a major boost. Those who had gained the most aerobic fitness from 15 to 18, as assessed from their high school phys ed marks, scored far better on the cognitive tests than those who had lost fitness. Since 268,496 of the subjects were brothers, the researchers were also able to determine that the links between aerobic fitness and intelligence were primarily due to environmental factors like exercise, rather than genetic factors.

The fact that aerobic exercise improves intelligence but strength training doesn't may come as a surprise. Researchers believe that many of exercise's neural benefits relate to whole-body effects such as increased blood flow: getting your heart pumping in a cardio workout carries more blood, along with helpful growth factors, to your brain. A 2009 study by University of North Carolina researchers used magnetic resonance angiograms to determine that elderly subjects who did regular aerobic exercise had more small blood vessels in their brains, and fewer twists and turns in those vessels, compared with non-exercising controls. The benefits of strength training, in contrast, tend to be limited to the muscles you're using.

Although it takes time to rewire your brain, you can tap into some of exercise's brain-boosting benefits almost instantly. In 2009, researchers at the University of Illinois at Urbana-Champaign put 21 volunteers through a set of tests to assess working memory (the ability to remember something and then retrieve it for use a short time later) immediately after a 30-minute session of either aerobic or resistance exercise, and then repeated the tests half an hour later. The aerobic exercisers improved their reaction time on the post-exercise test and improved it even more on the second test; the strength trainers, on the other hand, were no different from controls who hadn't exercised at all. These findings apply only to the specific working memory task that was tested, but they suggest that the mental benefits of aerobic exercise start right away.

There are also indications that more (or harder) exercise produces greater cognitive gains. Another 2009 study, from Taiwan's National Cheng Kung University, found that mice forced to run on a treadmill made greater cognitive gains than mice that ran at leisure on an exercise wheel (though both groups did improve). But there are limits. Extreme exertion like running a marathon generates stress hormone levels comparable to those seen in military interrogations and first-time parachute jumpers, which can interfere with some mental processes. Researchers tested 141 runners immediately after they completed the Boston or New York marathons and found that their "explicit memory," which answers questions like "What happened an hour ago?," was impaired. On the other hand, their "implicit memory," measured by the ability to complete partial words, was enhanced.

New results in this area continue to be published on a regular basis, so it won't be long before we're able to say with certainty why the extreme stress of a marathon helps some mental processes and hurts others, or which particular exercise-produced growth factors are key to generating new brain cells. For now, the advice is simple: keep doing all the exercise that's recommended for a healthy cardiovascular system, and you'll get a mental edge as a bonus.

CHEAT SHEET: MIND AND BODY

- Mental fatigue causes a reduction in physical performance, which suggests that exhaustion is controlled by the brain's perception of effort rather than the body's failure.
- The most productive training is "deliberate practice," which involves setting goals, monitoring progress, and focusing on technique rather than mindlessly repeating drills.
- Responses to music are highly personal, though there are some general patterns (faster music makes you work harder). Watching video is so distracting that it may lead you to slack off.
- Once you've mastered skills, whether it's golf putting or darts, focusing too much on the details can lead to choking.
- Swearing or imagining yourself doing something evil taps into feelings of aggression that enhance physical performance.
- Prolonged physical exercise causes the release of endorphins, which can lead to runner's high—and exercise addiction.
- Training with a group leads to greater endorphin production, which enhances pleasure and performance. About a third of people prefer working out alone.
- Exercise makes you smarter and improves your memory, starting immediately. Aerobic exercise is more effective than strength training, and the harder the better.

12: The Competitive Edge

WHEN ALLEN IVERSON WAS CRITICIZED for missing practice after the Philadelphia 76ers were eliminated from the National Basketball Association playoffs in 2002, he lashed out at reporters. "We're talking about practice," he said. "Not the game that I go out there and die for and play every game like it's my last, but we're talking about practice, man. How silly is that?" Although Iverson's attitude leaves something to be desired, it hints at one of the ingredients shared by great athletes: the ability to rise to the occasion in competition.

If you prepare for a game or race in the same way that you prepare for your usual workouts, you won't be well-rested enough to maximize your performance. On the other hand, you shouldn't try new foods, sleep habits, or training techniques right before a competition. By fine-tuning these elements and developing a familiar routine, you can give yourself an edge over your competitors when it counts and elevate your game (not practice).

How should I adjust my training in the final days before a competition?

In general, the fitter you are, the better you'll perform in any athletic contest. As a result, many beginners try to pack in as much training as possible in the final days before a race—hoping, perhaps, to make up for skipped sessions in previous months. But this is one kind of test you can't cram for. Every training session you do stresses your body, causing it to adapt in response—eventually. But this doesn't happen instantly, so sessions in the last week or two just add to your fatigue without boosting your fitness. It's not a good idea to stop training entirely, though, since you need to keep reinforcing the needed patterns of muscle movement. Instead, most athletes perform a "taper" in which training is gradually reduced, leading up to the competition in a way that maximizes fitness and minimizes fatigue.

The chief training variables that you can play with in endurance sports are volume, frequency, and intensity: how far you go, how often, and how hard. You can reduce your training all at once with a "step" taper or gradually cut back with a "progressive" taper. To evaluate these options, Laurent Bosquet of the University of Montreal and his colleagues assembled 182 studies on tapering in runners, swimmers, and cyclists and combined the most rigorous studies into one giant data set. The conclusion: the most effective taper involves a progressive reduction in training volume of 41 to 60 percent over a period of 8 to 14 days. You shouldn't change how often or how fast you do your workouts—just make them shorter.

There are a few caveats, Bosquet points out. First, the individual response to tapering can differ greatly: "Some athletes need as few as three or four days to dissipate fatigue, while others need three to four weeks," he says. This also depends on your level of training: if you're running only three times a week, your taper might simply consist of resting for the final two days before a race. A more subtle limitation is that studies tend to stick to tried-and-true tapering techniques, since it's hard to find enough competitive athletes willing to risk their season by trying a researcher's untested tapering scheme. For instance, theoretical models have predicted that resting in anticipation of a major competition, then ramping the training load back up again just before the race would actually be more effective than a straightforward taper. Bosquet is experimenting with this technique with athletes he coaches, but it has yet to be studied in a full-fledged trial.

Team sports are more difficult to analyze because success depends on so many different factors, and because high levels of performance need to be maintained over many weeks during playoff seasons. But the same principles of balancing the benefits of training with the risks of fatigue apply. Leading up to the 2002 World Cup, Swedish researchers recruited 11 top-notch European soccer teams (including Manchester United, Arsenal, AC Milan, Juventus, and Real Madrid) to monitor the training and playing load of their players. The data showed that the players who went on to underperform at the World Cup had played an average of 12.5 matches during the final 10 weeks of the season, while the players who exceeded expectations had played only nine matches—a clear sign that some form of tapering helped players get rid of fatigue and play their best on the biggest stage.

There's also a psychological component: reducing training makes some athletes feel anxious and lose confidence. The 2007 NCAA cross-country running champion, Josh McDougal of Liberty University, raised eyebrows by admitting that he had run a staggering 110 miles in the final week before his big win, a mere 10 percent reduction from his highest mileage. A more conventional taper the year before had produced a disappointing result: "Last year, I ran 48 miles the week of nationals, and my legs just felt terrible," he said after the race. "I just run well off training hard." That's why Bosquet recommends using the results of his analysis as a starting point, and then adjusting the parameters based on your experiences. Like McDougal, you'll eventually find a formula that gets you to the start line feeling both confident and well-rested.

Should I have sex the night before a competition?

The debate over whether pre-competition sex helps or hurts athletic performance tends to be argued with clichés rather than scientific studies. The conventional wisdom was articulated by Mickey Goldmill, the hard-nosed trainer in the original *Rocky* movie: "Women weaken legs." Legendary New York Yankees manager Casey Stengel took a more conciliatory stance: "It's not the sex that wrecks these guys, it's staying up all night looking for it."

If you're looking for advice that has been tested in the lab, your options are much more limited. Samantha McGlone and Ian Shrier of McGill University found only three relevant studies of morning-after prowess in a comprehensive review of the literature published in the *Clinical Journal of Sport Medicine* in 2000. One measured grip strength in married male athletes, either after sex the night before or after at least six days of abstinence, and found no difference. A similar study at Colorado State University looked at a wider range of indicators including reaction time, stair-climbing, and balance, again with no apparent effect. Finally, a treadmill test of subjects who had been randomly assigned to either have sex or abstain 12 hours before the test found no effect on aerobic power and two other variables.

Given that a "normal" bout of sex burns energy equivalent to climbing just two flights of stairs, the lack of effect is not surprising. The review also exposes some major gaps in our knowledge. For instance, the studies were only on men—the question of whether women's legs are weakened doesn't appear to even arise in the literature. And those gaps haven't been filled in the decade since the review appeared, Shrier says.

Physiological changes are only part of the equation, though. Sex undoubtedly affects mood, and changes in traits such as aggression could influence performance. But sports psychologists believe that the optimal state of mental arousal is very personal. Some athletes need to be psyched up, and others need to be calmed down in order to perform at their best. That means that there likely isn't a definitive answer that applies to everyone—which leaves the final word to McGlone, who went on to be the top Canadian triathlete at the 2004 Olympics and is now one of the top Ironman triathletes in the world. "All I can advise is, before a big race, stick with your usual routine, whatever that may be," she says, adding a note of Stengel-esque caution: "Just try to get a good night's sleep."

Can drinking slushies boost my performance on hot days?

When you exercise in the heat, your core temperature rises until it hits a critical value—typically around 104°F (40°C) or a little lower—and you're forced to stop. One way to delay this moment is to "precool" your body, an approach pioneered by Australian sports scientists. It was the Australians who showed up at the Atlanta Olympics in 1996 with form-fitting ice vests for their endurance athletes (which you can now buy for around $200), and they followed up at the 2004 Athens Olympics with ice baths for their athletes to plunge in immediately before competing.

Although these methods seem to work, they're a little too unwieldy to become very widespread. But for the 2008 Olympics in Beijing, the Aussies unveiled a new secret weapon with truly mass-market appeal: the slushie. As Louise Burke, the head of sports nutrition at the Australian Institute of Sport, explained at a conference shortly after the games, ingesting a crushed-ice drink cools the athlete internally—not just with the frigid temperature, but due to the additional "phase change" energy required to melt the ice from solid to liquid.

In pre-Olympic tests, researchers found that the athletes lowered their internal temperature by more than 1°F by drinking the slushies, which consisted of 14 mL per kilogram of body weight of a half-and-half mix of sports drink and water. This translated into an advantage of about one minute in a 79-minute cycling time trial compared to controls. Plunging into a cold bath at 54°F (12°C) for 10 minutes cooled the cyclists even more—so much, in fact, that they started the time trial too fast and faded in the latter stages. That left the slushie as the best option.

Further research by scientists at Edith Cowan University in Western Australia, published in 2010, has confirmed the benefits of "ice slurries." In this study, the subjects drank either a slushie at 30°F (-1°C) or cold water at 39°F (4°C). Thanks to the phase change energy, the slushie group decreased their rectal temperature by 1.19°F (0.66°C) and lasted 50.2 minutes in a cycle test to exhaustion in the heat, while the cold water group cooled by just 0.45°F (0.25°C) and lasted only 40.7 minutes.

Interestingly, the slushie group managed to keep biking until their core temperature reached 39.36°C, while the cold water group was forced to stop at 39.05°C—a small but significant difference. The researchers speculate that the subjects may have cooled their brains slightly as the slushies passed through their mouth and throat. Since the decision to terminate exercise in the heat is thought to be controlled centrally, a cooler brain may have permitted the rest of the body to get a little hotter than usual before calling a halt. Earlier studies with dogs and goats have suggested that brain temperature, rather than core temperature, might control the limit of exercise tolerance in the heat.

The performance boost offered by slushies, ultimately, is a few percent at most. But it's a simple intervention that could easily be implemented at big sporting events—far more easily than ice baths, and far more cheaply than cooling vests. That's one reason the Australians brought seven slushie machines to Beijing and used them for soccer, track, cycling, triathlon, rowing, field hockey, and several other sports. The other reason is familiar to every athlete looking for a slight edge wherever he or she can find it: the mental boost. "In Beijing, we wanted something new," Burke admitted. "You always have to have something new for athletes for that placebo effect."

Will drinking coffee help or hinder my performance?

Until 2004, Olympic athletes could test positive for caffeine if they drank as few as three cups of strong coffee. Then, frustrated with trying to regulate such a commonly used substance, the World Anti-Doping Agency (WADA) removed caffeine from its list of restricted substances—and the strangest thing happened. After the ban was lifted, caffeine levels found in WADA urine tests actually decreased in almost all sports. If it wasn't worth banning, athletes apparently figured, it wasn't worth taking.

They were wrong.

"I don't think there's any doubt that caffeine is a very powerful ergogenic [performance enhancing] aid," says University of Guelph professor Terry Graham, one of the world's leading researchers on the topic. "It's probably the most versatile aid out there." After decades of studies, it's now well established that caffeine helps sprint performance and improves endurance in activities lasting up to two hours. There's also increasingly solid evidence that it helps resistance exercise like weightlifting.

The usual counterargument is that caffeine's diuretic effect can leave you dehydrated, ultimately hurting performance, particularly in endurance events. But recent research has thoroughly dispelled that notion, Graham says. Another commonly repeated myth that has now been disproved is that caffeine's performance boost results from the body burning more fat for energy, he says.

How caffeine does work is still up for debate. Caffeine is a stimulant, and it may also carry a placebo effect for some athletes—but that's not the whole story either. "If I were to place electrodes on your muscle, and start to stimulate it so that your muscle is contracting and your brain's not involved, I can still see an effect [from the caffeine]," Graham says. The current theory is that caffeine directly affects how muscle fibers contract at a cellular level, making each fiber contract more strongly when it receives a signal from the nervous system.

Caffeine, however, is not the same thing as coffee. The only rigorous study directly comparing the effects of caffeine (in pill form) with coffee was performed in Graham's lab. Surprisingly, he and his colleagues found that only pure caffeine produced a performance boost, even when the level of caffeine in the bloodstream from coffee was identical. "We didn't believe it at first, so we kept adding subjects," he says, "but the data just got stronger." Other studies have found a performance-enhancing effect from coffee, so Graham is cautious about overstating his results. What is clear is that the effects of coffee, with its complex mix of bioactive ingredients, are far harder to nail down than the unambiguous effects of pure caffeine.

Current estimates suggest that somewhere between 82 and 92 percent of North American adults consume coffee on a regular basis. The use of caffeine explicitly for its performance-enhancing qualities is also widespread: a British study published in 2008 found that 60 percent of cyclists and 33 percent of track and field athletes took caffeine to enhance athletic performance. International-caliber track athletes were twice as likely to seek a boost from caffeine as club-level athletes. But Graham sounds a note of caution about the value of a caffeine-fueled personal best for recreational athletes, who are typically focused on beating their own best performances rather than beating specific competitors. In effect, it simply moves the finish line closer. So despite all his research, Graham has never used caffeine in 20 years of running marathons.

Does competing in front of a crowd improve performance?

One of the unique aspects of running is that ordinary weekend warriors have the opportunity to toe the starting line right next to the fastest runners in the world and run through the streets of big cities with thousands of spectators cheering them on. In fact, some major races like the Marine Corps Marathon in Washington, D.C., draw big crowds of spectators without any elite runners at all. Not surprisingly, many runners aim to set their personal best times at these races, reasoning that the energy of the crowd will help propel them forward. But does it really make a difference?

Researchers have been studying the effect of spectators on sports performance for years, in an attempt to understand the well-known "home advantage" experienced by professional sports teams. This effect is thought to stem from a wide variety of factors such as biased officials, travel-weary opponents, and elevated testosterone in home-team players eager to protect their home turf. While you'd think the crowd would be a key part of home advantage, a surprising study published in 2010 by Niels van de Ven of Tilburg University in the Netherlands shows that this isn't necessarily the case.

Van de Ven took advantage of two quirks of the schedules of Italian soccer leagues during the 2006–2007 seasons. First, he examined 20 games that were played in empty stadiums, since the home teams were being penalized for misbehaving fans and safety infractions. After carefully controlling for the quality of the teams, he found that the same home advantage existed with or without a crowd. Second, he examined games between teams that shared the same home stadium, like AC Milan and Internazionale or AS Roma and Lazio Roma. In these cases, the home team has dominant crowd support because of the presence of its season-ticket holders. However, no home advantage could be detected in these games, suggesting that home advantage (in soccer, at least) likely results from familiarity with the stadium rather than from loud cheering.

This doesn't mean that you won't get a performance boost from lining up at a big race—but it suggests that the encouraging shouts from friends, family, and complete strangers aren't the crucial factor. Instead, it's the adrenaline-producing stress and anxiety of participating in a big event that harnesses your fight-or-flight instincts and allows you to exceed your usual abilities. "For your local marathon, you will not be as pumped up as for the Boston Marathon," van de Ven says. But there's a trade-off, as the study suggests: if you run your local marathon, you'll not only benefit from sleeping in your own bed and avoiding travel, but you may also be able to train on the course to become familiar with it. The best choice may depend on your personality: if you're usually anxious before races, the local race may be your best bet; if you need help getting psyched up, taking a trip to a big event could give you a boost.

How much sleep do I need for optimal physical performance?

For top athletes, getting enough sleep has long been considered the sort of bland good advice that is obvious but easy to ignore—like eating lots of vegetables. A pair of recent pilot studies by Charles Samuels, the medical director of the Centre for Sleep and Human Performance in Calgary, confirms that poor sleep quality is prevalent even in Olympic-level athletes (in this case from the national bobsleigh and skeleton teams). But the problem is even worse for ordinary people: "It's average athletes who are the most likely to curtail their sleep to train," Samuels says. "They're getting up at 4 a.m. to run for an hour so they can get to work by 7 a.m."

That's not necessarily a winning strategy, especially for people who are already operating on the least amount of sleep that they can handle. Incurring a steadily mounting sleep debt has well-known effects on mood and cognitive ability, and a few studies are now suggesting that sleep also has direct links with physical performance. For example, Stanford University sleep researcher Cheri Mah has conducted a series of small studies testing athletes on the university's teams. When five varsity swimmers increased their sleep time to ten hours a night from their typical six to nine hours, they slashed 0.15 seconds from their reaction time off the start and similarly improved their turn time, 15-meter sprint time, and kick rate. Similarly, increased sleep improved sprint time and free-throw percentage for a group of basketball players.

With only a few participants and no control group, these results are far from definitive, but they represent a first step to quantifying the athletic benefits of sleep. Samuels, meanwhile, has been working on a project with the Canadian downhill ski team to investigate the link between inadequate sleep and injuries, as well as studies of how globe-trotting athletes can best adjust to crossing time zones.

It's still a challenge to apply these results in the real world. "I know it sounds ridiculous to get 10 hours of sleep a night," Mah admits. "That's an extreme." For the typical person, she says, consistently increasing the amount of nightly sleep by even a small amount can produce positive effects. Most adults need seven to eight hours of sleep nightly, while teens and young adults need nine or more, though there's quite a bit of individual variation. One of the most interesting implications of her studies with varsity athletes is that even just a few weeks of concerted sleep catch-up has a measurable effect on performance—something to keep in mind before the next big game or race.

It's also worth noting that, just as sleep helps exercise, the converse is also true. A 2010 study from the Federal University of São Paulo found that moderate aerobic exercise (but not strength training or heavy aerobic exercise) increased reported sleep time by 26 percent in a group of chronic insomniacs. One caveat, notes Samuels, is that exercise in the three hours before bedtime can actually hinder sleep in adults in their 30s and older—if they already struggle with sleep. "If you're a good sleeper," he adds, "nothing matters."

How should I pace myself in a long-distance race?

The usual advice, lifted straight from Aesop's Fables, is that slow and steady wins the race. This is a prudent approach, especially for inexperienced racers, whether you're running, biking, swimming, skating, rowing, snowshoeing, or undertaking any other activity where you're hoping to reach the finish line at precisely the moment that you expend your last ounce of energy. But for those who have already run a few races and are looking to improve their time, research suggests a higher-risk, higher-reward approach.

Exercise physiologist Robert Kenefick and his colleagues at the University of New Hampshire tested various pacing strategies for the most popular road race distance, 5K, in a 2006 study. He had test subjects—who were serious recreational runners but not elite athletes—run a series of races with the speed of the first mile carefully controlled, either at an even pace based on their best time, or 3 or 6 percent faster. To everyone's surprise, the fastest overall times came from the fastest opening mile, while the slowest races came from running an even pace. The researchers measured physiological variables such as oxygen use and heart rate while the runners ran, but they couldn't detect any difference between the paces.

This finding is consistent with observations of the fastest runners in the world. Ross Tucker and his colleagues at the University of Cape Town in South Africa analyzed every men's world record ever set over 5,000 and 10,000 meters on the track for a study in the *International Journal of Sports Physiology and Performance*. The pattern is remarkably consistent: a fast start, gradually slowing during the middle of the race, then a fast finish. In fact, in 63 of the 64 world records they studied, the first and last kilometers were faster than any other kilometer in the race. The only exception was Paul Tergat's 1997 record over 10,000 meters, in which the ninth kilometer was one second faster than the tenth kilometer.

This fast finish isn't necessarily a deliberate strategy. Rather, Tucker argues that it reflects "anticipatory regulation" of effort, in which your conscious and unconscious minds work together so that you reach the finish line having worked as hard as possible, but maintaining a reserve to ensure that you don't collapse prematurely. This is why prior racing experience is essential for an aggressive pacing strategy: your brain needs some basis of comparison to figure out when the end will come.

Exercise scientists sometimes refer to the effects of knowing where the finish line is as "teleoanticipation." Studies over the course of several decades have shown how powerful these effects can be. For example, researchers in a 1980 study told one group of volunteers to run on a treadmill for 20 minutes and told another group to run at the same pace for 30 minutes—but stopped both groups after 20 minutes. Even though the physical demands were identical, the subjects who thought they would have to run for another 10 minutes reported much lower ratings of perceived exertion, showing that our feelings of fatigue are linked to when we expect to finish.

These findings don't mean that you should sprint all-out off the starting line and simply hope to hang on. The best performers in Tucker's study were those who started fast but then managed to settle into a steady pace, rather than getting steadily slower throughout the race. So be cautious at the start—but don't be afraid to push a little faster than your expected final pace. "Either way, you're tired at the end," says Kenefick, who is now a U.S. Army research physiologist. "So if you go out too slow, you can't make that up at the end."

Is endurance or sprint speed more important in field sports like soccer?

For decades, sports scientists have focused their attention on two opposite ends of the athletic spectrum: endurance and speed. But the demands of sports like soccer, rugby, and basketball don't fit neatly into either one of these pigeonholes. Time–motion studies using GPS and video tracking have found that top soccer players cover up to 7.5 miles over the course of a typical match, which certainly requires endurance. On the other hand, games are won and lost when one player beats another to the ball by a few inches. The same analyses find that players make somewhere between 20 and 60 short, all-out sprints during a match, each lasting two to four seconds and covering 10 to 30 yards. It's this "repeated sprint ability" that separates good players from great ones, and scientists have finally started studying how to improve this trait.

The first time you sprint during a game, about 80 percent of the energy you need is provided by short-term fuel sources that don't require any oxygen, while the remaining 20 percent is aerobic. With short recoveries between sprints, the aerobic component rises to about 50 percent when you make your third sprint, and eventually reaches about 75 percent for each sprint after you've been playing for a while. At this point, according to McMaster University researcher Stuart Phillips, you're relying on carbohydrate stores in the same way long-distance runners do. As a result, the optimal fueling strategies are essentially the same as for endurance challenges: make sure your carbohydrate stores are full before the game starts (see p. 206), and keep topping them up throughout.

So what's the best way to improve your ability to keep sprinting late in the game? In a 2010 study, French researchers compared two approaches. One group of elite teen soccer play-ers focused on developing their explosive strength once a week, performing a series of drills, including vertical and horizontal jumps, hurdles, and sprints. Another group focused on repeated sprint training, doing up to three sets of six shuttle sprints (run-ning 20 meters, touching the ground, then returning to the start position as quickly as possible) with about 20 seconds of rest be-tween sprints. Both groups improved the maximum speed for a single sprint, but only the second group improved their times for repeated sprints—a crucial distinction, since several studies have found that performance on repeated sprint tests is a strong predictor of better performance in matches.

Similar patterns are seen in other sports. Rugby is similar to soccer, with players covering about six miles and spending about 25 percent of that time in the "critical performance zone" where they're chasing a ball or an opponent at close to maximum inten-sity. In ice hockey, players cover only about 2.5 miles but spend half that time in the critical performance zone. Basketball play-ers, competing on a smaller surface, cover about 1.3 miles and spend 20 percent of that time in the critical performance zone. Each sport has slightly different demands, but they all have a start-stop rhythm that you can prepare for by doing repeated sprints with short rests and rapid changes of direction.

CHEAT SHEET: THE COMPETITIVE EDGE

- Gradually reduce your training volume by 41 to 60 percent over a period of 8 to 14 days before a competition to maximize performance. Don't change training frequency or intensity.
- Sex before competition is unlikely to have any physical effects but could affect mental readiness. Stick to a familiar routine.
- Downing a slushy ice beverage can lower your core temperature enough to boost endurance on hot days.
- Caffeine is a powerful performance enhancer, acting as a stimulant and directly on your muscles; coffee has less predictable effects due to its complex mix of ingredients.
- Competing in a familiar environment may offer greater advantages than having a large crowd cheering you on, though individual responses vary.
- Getting enough sleep boosts performance; even just a few weeks of concentrated sleep catch-up has measurable effects on speed and reaction time.
- Contrary to the usual advice, a slightly faster start may help you finish with a quicker time than a perfectly even pace in long races.
- To train for the mix of speed and endurance required in field sports like soccer, practice repeated shuttle sprints (20 meters) with sharp turns and short recoveries (20 seconds).

Conclusions: From Lab to Gym

AFTER A WHIRLWIND TOUR, YOU'RE UP TO DATE on the latest thinking from scientists around the world about the merits of personal trainers, probiotics, Pilates, plasma-rich platelets, and a host of other topics. "Now you know," as the public service announcements at the end of G.I. Joe cartoons used to say, "and knowing is half the battle." The other half is the real challenge: putting this knowledge into practice. To that end, I hope you'll take the following three basic messages from this book:

1. **Do something rather than nothing.** You might be daunted by the sheer number of exercise choices described. But there's no wrong answer—pick something you enjoy. You might also be daunted by how challenging some of the exercise programs sound, and think that you'll never be a marathoner or a weightlifter or a mountain hiker. But if there's one overriding theme in the research presented here, it's that any exercise, in almost any amount, brings significant and immediate health benefits. Start doing it, and worry about getting it right later.
2. **Figure out your goals and monitor your progress.** One of the reasons this book couldn't be condensed into a five-page pamphlet is that everyone has different goals. The workout routine that's perfect for your sister might make

no sense for you. Think carefully about what you hope to achieve in six months, a year, five years—bearing in mind the aphorism that most people overestimate what they can achieve in the short term and underestimate what they can achieve over the long term. Choose a program that will move you toward those goals, and monitor your progress, whether it's how far you can walk, how much weight you can lift, how well you can serve a tennis ball, or even how you feel. If you don't start to see progress after 6 to 12 months, consider whether your program is appropriate to your goals.

3. **Try something new.** Whenever researchers line up two or more exercise techniques against each other, the conclusion is almost never "A is better than B," or "A and B are the same." Instead, it's "A has these strengths and weaknesses, while B has these other strengths and weaknesses." Moreover, all programs suffer from diminishing returns after a few years—if you always bike at the same pace and do the same five strength exercises, your improvements will be measured in fractions of a percent. Trying something new every now and then will force your body to adapt in new ways, and keep you mentally fresh.

Of course, scientists are still learning more about the workings of the body, churning out hundreds of papers a month in journals like *Medicine & Science in Sports & Exercise,* the *Journal of Strength and Conditioning Research,* the *British Journal of Sports Medicine,* and many others. Most of these papers are building on existing knowledge, filling in gaps, and adding details.

But sometimes there are radical new findings. Who would have guessed, a decade ago, that experts would be dismissing the injury-prevention benefits of static stretching and celebrating lactate as a crucial fuel for exercise? Unfortunately, health (and science in general) is often reported in the media as a series of unconnected breakthroughs, with big headlines and no context. When it seems that every study you read about is contradicted by another one a few months later, it's hard to figure out what to believe. Look for high-quality reporting that explains how the study was done, who paid for it, and how it relates to previous findings in the area. Gina Kolata and Gretchen Reynolds in the *New York Times* and Amby Burfoot in *Runner's World* are good examples of journalists who decode scientific finding in an accessible, hype-free manner.

With these principles in mind, you have the tools you need to develop and follow a well-rounded exercise plan. Be patient, consistent, and (with the help of endorphins, perhaps) enjoy it!

Acknowledgments

In February 2008, I was discussing the possibility of writing a regular column on the science of exercise with Kevin Siu, then the deputy editor of the *Globe and Mail*'s Life section. "Do you think you'd be able to find enough material to keep a column going for a year or two?" Kevin asked. No problem, I replied. There's enough to fill a book. Since then, Kevin and his successor as handling editor of the column, Cliff Lee, have played crucial roles in guiding my research, generating new ideas, and focusing my writing. Without them and their colleagues at the *Globe*—particularly the eagle-eyed and indefatigable copy editors—this book wouldn't exist.

One of the best parts about working with the *Globe* was the illustrator assigned to the column, Trish McAlaster. With a degree in phys ed and past experience as both a personal trainer and a medical illustrator, Trish couldn't have been a more ideal partner to help express complex training exercises and scientific concepts in an accessible way. I was thrilled when she agreed to illustrate the book and am even more thrilled with the result.

The book itself began to take shape under the expert guidance of my agent, Rick Broadhead. Its current scope and format owe a lot to Rick's insights, and I appreciate how patient and responsive he was throughout the process. The fact that I ended up with McClelland & Stewart wasn't just a business decision:

I'm proud to be associated with such an august Canadian institution, and I appreciate the faith that Jenny Bradshaw, my editor there, showed in me even before she saw my proposal. Jenny and Stephanie Meyers, her counterpart at HarperCollins in the United States, formed a dynamic editing duo who made the book unequivocally better. I'm grateful for all their suggestions, from changing a comma to adding a chapter.

My biggest debt is to the hundreds of scientists whose work is described in these pages—not only for their research, but for sharing their time and in some cases welcoming me into their labs. Some, like Trent Stellingwerff, Bradley Young, Stuart Phillips, Martin Gibala, and Carl Foster, have helped me on multiple occasions. Trent, in particular, has been an exceptional resource in the area of sports nutrition and agreed to read an early draft of the manuscript for me.

I'd also like to thank Michal Kapral and the staff at *Canadian Running*, as well as my editors at *The Walrus* and *Runner's World*, where I first wrote about some of the research presented here.

Finally, I'd like to thank my family. It's not possible to express here the debt of gratitude I owe my parents, Moira and Roger, for all their support; suffice to say that housing and feeding me—as they did for several extended stints while I was writing this book—was the least of their contributions. My brother, Tim, was the first beta-tester of the book's content during his inspirational and highly successful journey to fitness after a serious bike accident. And my wife, Lauren, has been a constant partner in this process, asking new questions and answering other ones by drawing on her experiences as an athlete, kinesiologist, and medical student. I'm lucky to have someone who shares my passions in life and also makes sure that I eat my vegetables and do some push-ups.

References

Introduction

F. W. Booth and M. J. Laye, "Lack of adequate appreciation of physical exercise's complexities can pre-empt appropriate design and interpretation in scientific discovery," *Journal of Physiology*, 2009, 587(23), 5527–5540.

John Hawley and John Holloszy, "Exercise: It's the real thing!," *Nutrition Reviews*, 2009, 67(3), 172–178.

Chapter 1: Getting Started

How long does it take to get in shape?

Megan Anderson et al., "Training vs. body image: Does training improve subjective appearance ratings?" *Journal of Strength and Conditioning Research*, 2004, 18(2), 255–259.

Vernon Coffey and John Hawley, "The molecular bases of training adaptation," *Sports Medicine*, 2007, 37(9), 737–763.

J. A. Hawley and S. J. Lessard, "Exercise training-induced improvements in insulin action," *Acta Physiologica*, 2008, 192, 127–135.

Keitaro Kubo et al., "Time course of changes in muscle and tendon properties during strength training and detraining," *Journal of Strength and Conditioning Research*, 2010, 24(2), 322–331.

Am I exercising enough?

William Haskell et al., "Physical activity: Health outcomes and importance for public health policy," *Preventive Medicine*, 2009, 49, 280–282.

Michael Leitzmann et al., "Physical activity recommendations and decreased risk of mortality," *Archives of Internal Medicine,* 2007, 167(22), 2453–2460.

Paul Williams, "Relationship of incident glaucoma versus physical activity and fitness in male runners," 2009, *Medicine & Science in Sports & Exercise,* 2009, 41(8), 1566–1572.

Paul Williams, "Usefulness of cardiorespiratory fitness to predict coronary heart disease risk independent of physical activity," *American Journal of Cardiology,* 2010, 106(2), 210–215.

What should I do first: cardio or weights?

G. A. Nader, "Concurrent strength and endurance training: From molecules to man," *Medicine & Science in Sports & Exercise,* 2006, 38(11), 1965–1970.

Can I get fit in seven minutes a week?

Martin Gibala et al., "Short-term sprint interval training versus traditional endurance training: Similar initial adaptations in human skeletal muscle and exercise performance," *Journal of Physiology,* 2006, 575, 901–911.

R. E. Macpherson et al., "Run sprint interval training improves aerobic performance but not max cardiac output," *Medicine & Science in Sports & Exercise,* 2010, published online ahead of print.

Jason Talanian et al., "Two weeks of high-intensity aerobic interval training increases the capacity for fat oxidation during exercise in women," *Journal of Applied Physiology,* 2007, 102, 1439–1447.

Can exercise increase my risk of a heart attack?

Barry Maron, "Sudden death in young athletes," *New England Journal of Medicine,* 2003, 349, 1064–1075.

Donald Redelmeier and Ari Greenwald, "Competing risks of mortality with marathons: Retrospective analysis," *BMJ,* 2007, 335, 1275–1277.

Will exercising in cold air freeze my lungs?

John Castellani et al., "Prevention of cold injuries during exercise," American College of Sports Medicine position stand, 2006.

Tina Evans et al., "Airway narrowing measured by spirometry and impulse oscillometry following room temperature and cold temperature exercise," *Chest*, 2005, 128, 2412–2419.

When is it too hot to exercise?

Marius Brazaitis et al., "The effect of two kinds of T-shirts on physiological and psychological thermal responses during exercise and recovery," *Applied Ergonomics*, 2010, published online ahead of print.

William Roberts, "Determining a 'Do Not Start' temperature for a marathon on the basis of adverse outcomes," *Medicine & Science in Sports & Exercise*, 2010, 42(2), 226–232.

Nigel Taylor, "Immersion treatment for exertional hyperthermia: Cold or temperate water?" *Medicine & Science in Sports & Exercise*, 2010, 42(7), 1246–1252.

Ross Tucker et al., "Impaired exercise performance in the heat is associated with an anticipatory reduction in skeletal muscle recruitment," *Pflügers Archiv—European Journal of Physiology*, 2004, 448(4), 422–430.

Should I avoid exercising outside when air pollution is high?

Chauncy Blair et al., "Volatile organic compounds in runners near a roadway: Increased blood levels after short-duration exercise," *British Journal of Sports Medicine*, 2010, 44(10), 731–735.

Jeroen Johan de Hartog et al., "Do the health benefits of cycling outweigh the risks?" *Environmental Health Perspectives*, 2010, 118(8), 1109–1116.

R. T. O'Donoghue et al., "Exposure to hydrocarbon concentrations while commuting or exercising in Dublin," *Environment International*, 2007, 33, 1–8.

Jette Rank et al., "Differences in cyclists and car drivers exposure to air pollution from traffic in the city of Copenhagen," *The Science of the Total Environment*, 2001, 279, 131–136.

Kenneth Rundell, "Vehicular air pollution, playgrounds, and youth athletic fields," *Inhalation Toxicology*, 2006, 18, 541–547.

How will exercise affect my immune system?

Marian Kohut et al., "Chronic exercise reduces illness severity, decreases viral load, and results in greater anti-inflammatory effects than acute exercise during influenza infection," *Journal of Infectious Diseases*, 2009, 200, 1434–1442.

T. Lowder et al., "Moderate exercise protects mice from death due to influenza virus," *Brain, Behavior and Immunity*, 2005, 19(5), 377–380.

Stephen Martin et al., "Exercise and respiratory tract viral infections," *Exercise and Sport Sciences Reviews*, 2009, 37(4), 157–164.

C. E. Matthews et al., "Moderate to vigorous physical activity and risk of upper-respiratory tract infection," *Medicine & Science in Sports & Exercise*, 2002, 34(8), 1242–1248.

David Nieman, "Marathon training and immune function," *Sports Medicine*, 2007, 37(4–5), 412–415.

Is motivation to exercise genetic?

Maria Åberg et al., "Cardiovascular fitness is associated with cognition in young adulthood," *PNAS*, 2009, 106(49), 20906–20911.

Marleen de Moor et al., "Genome-wide association study of exercise behavior in Dutch and American adults," *Medicine & Science in Sports & Exercise*, 2009, published online ahead of print.

J. H. Stubbe et al., "Genetic influences on exercise participation: A comparative study in adult twin samples from seven countries," *PLoS One*, 2006, 1(1).

How long does it take to get unfit?

R. C. Hickson, "Reduced training intensities and loss of aerobic power, endurance, and cardiac growth," *Journal of Applied Physiology*, 1985, 58(2), 492–499.

Keitaro Kubo et al., "Time course of changes in muscle and tendon properties during strength training and detraining," *Journal of Strength and Conditioning Research*, 2010, 24(2), 322–331.

Iñigo Mujika and Sabino Padilla, "Cardiorespiratory and meta-bolic characteristics of detraining in humans," *Medicine & Science in Sports & Exercise*, 2001, 33(3), 413–421.

Rasmus Olsen et al., "Metabolic responses to reduced daily steps in healthy nonexercising men," *Journal of the American Medical Association*, 2008, 299, 1261–1263.

Paul Williams, "Asymmetric weight gain and loss from increasing and decreasing exercise," *Medicine & Science in Sports & Exercise*, 2008, 40(2), 296–302.

Chapter 2: Fitness Gear

Is running on a treadmill better or worse than running outside?

Andrew Jones and Jonathan Doust, "A 1% treadmill grade most accurately reflects the energetic cost of outdoor running," *Journal of Sports Sciences*, 1996, 14(4), 321–327.

Patrick Riley et al., "A kinematics and kinetic comparison of over-ground and treadmill running," *Medicine & Science in Sports & Exercise*, 2008, 40(6), 1093–1100.

Is the elliptical machine just as good as running?

Michael Desch et al., "Cardiometabolic comparison of elliptical and treadmill exercise responses," *Medicine & Science in Sports & Exercise*, 2005, 37(5), S105.

M. Egaña and B. Donne, "Physiological changes following a 12 week gym based stair-climbing, elliptical trainer and treadmill running program in females," *Journal of Sports Medicine and Physical Fitness*, 44(2), 141–146.

Nikki Hughes et al., "Ratings of perceived exertion (RPE) during elliptical trainer, treadmill and recumbent bike exercise," *Medicine & Science in Sports & Exercise*, 2005, 37(5), S103.

Kathleen Knutzen et al., "Comparison of quadriceps femoris activation during elliptical and treadmill exercise in athletes with patellofemoral pain," *Medicine & Science in Sports & Exercise*, 2006, 38(5), S498.

Constance Mier and Yuri Feito, "Metabolic cost of stride rate, resistance, and combined use of arms and legs on the elliptical trainer," *Research Quarterly for Exercise and Sport*, 2006, 77(4), 507–513.

Do I really need specialized shoes for walking, running, tennis, basketball, and so on?

H. H. G. Handoll et al., "Interventions for preventing ankle ligament injuries," Cochrane Library, 2008.

Pui Wah Kong et al., "Running in new and worn shoes—a comparison of three types of cushioning footwear," *British Journal of Sports Medicine*, 2009, 8(1), 52–59.

Kai Way Li et al., "Physiological and psychophysical responses in handling maximum acceptable weights under different footwear-floor friction conditions," *Applied Ergonomics*, 2007, 38, 259–265.

M. S. Orendurff et al., "Regional foot pressure during running, cutting, jumping, and landing," *American Journal of Sports Medicine*, 2008, 36(3), 566–571.

Michael Ryan et al., "The effect of three different levels of footwear stability on pain outcomes in women runners: A randomised control trial," *British Journal of Sports Medicine*, 2010, published online ahead of print.

J. E. Taunton et al., "A prospective study of running injuries: The Vancouver Sun Run 'In Training' clinics," *British Journal of Sports Medicine*, 2003, 37, 239–244.

Will running barefoot help me avoid injuries?

D. C. Kerrigan et al., "The effect of running shoes on lower extremity joint torques," *PM&R*, 2009, 1(12), 1058–1063.

Daniel Lieberman et al., "Foot strike patterns and collision forces
in habitually barefoot versus shod runners," *Nature*, 2010, 463,
531–535.

Craig Richards et al., "Is your prescription of distance running
shoes evidence-based?" *British Journal of Sports Medicine*, 2009,
43(3), 159–162.

Will compression clothing help me exercise?

A. Bringard et al., "Compression élastique externe et fonction
musculaire chez l'homme," *Science & Sports*, 2007, 22, 3–13.

Duncan French et al., "The effects of contrast bathing and compression therapy on muscular performance," *Medicine &
Science in Sports & Exercise*, 2008, 40(7), 1298–1307.

Wolfgang Kemmler et al., "Effect of compression stockings on
running performance in men runners," *Journal of Strength and
Conditioning Research*, 2009, 23(1), 101–105.

William Kraemer et al., "Influence of compression garments on
vertical jump performance in NCAA Division I volleyball players," *Journal of Strength and Conditioning Research*, 1996, 10(3),
180–183.

Aaron Scanlan et al., "The effects of wearing lower-body compression garments during endurance cycling," *International Journal
of Sports Physiology and Performance*, 2008, 3(4), 424–438.

Does walking with poles give me a better workout?

H. Figard-Fabre et al., "Physiological and perceptual responses to
Nordic walking in obese middle-aged women in comparison
with the normal walk," *European Journal of Applied Physiology*,
2010, 108(6), 1141–1151.

Ernst Hansen and Gerald Smith, "Energy expenditure and comfort during Nordic walking with different pole lengths,"
Journal of Strength and Conditioning, 2009, 23(4), 1187–1194.

Thorsten Schiffer et al., "Energy cost and pole forces during
    Nordic walking under different surface conditions," *Medicine
    & Science in Sports & Exercise,* 2009, 41(3), 663–668.

Are sports video games real workouts?
Julio Bonis, "Acute Wiiitis," *New England Journal of Medicine,* 2007,
    356(23), 2431–2432.
Amanda Daley, "Can exergaming contribute to improving
    physical activity levels and health outcomes in children?"
    Pediatrics, 2009, 124, 763–771.
Karen Eley, "A Wii fracture," *New England Journal of Medicine,*
    2010, 362(5), 473–474.
L. Graves et al., "Comparison of energy expenditure in ado-
    lescents when playing new generation and sedentary com-
    puter games: Cross sectional study," *BMJ,* 2007, 335(7633),
    1282–1284.
Scott Leatherdale et al., "Energy expenditure while playing active
    and inactive video games," *American Journal of Health Behavior,*
    2010, 34(1), 31–35.

What should I do with wobble boards and exercise balls?
Michael Wahl and David Behm, "Not all instability training de-
    vices enhance muscle activation in highly resistance-trained
    individuals," *Journal of Strength and Conditioning Research,* 2008,
    22(4), 1360–1370.
Carolyn Emery et al., "A prevention strategy to reduce the inci-
    dence of injury in high school basketball: A cluster randomized
    controlled trial," *Clinical Journal of Sports Medicine,* 2007, 17(1),
    17–24.
Con Hrysomallis, "Relationship between balance ability, training
    and sports injury risk," *Sports Medicine,* 2007, 37(6), 547–556.
K. Söderman et al., "Balance board training: Prevention of trau-
    matic injuries of the lower extremities in female soccer play-
    ers?" *Knee Surgery, Sports Traumatology, Arthroscopy,* 2000, 8(6),
    356–363.

Evert Verhagen et al., "The effect of a proprioceptive balance board training program for the prevention of ankle sprains," *American Journal of Sports Medicine*, 2004, 32, 1385–1393.

Can a mouthpiece make me stronger, faster, and more flexible?
Dena Garner and Erica McDivitt, "Effects of mouthpiece use on airway openings and lactate levels in healthy college males," *Compendium of Continuing Education in Dentistry*, 2010, 31(6).
Dena Garner and Erica McDivitt, "The effects of mouthpiece use on salivary cortisol levels during exercise," *Medicine & Science in Sports & Exercise*, 2008, 40(5), S468.

Is there any benefit to strengthening my breathing muscles?
Andrew Edwards and Raewyn Walker, "Inspiratory muscle training and endurance: A central metabolic control perspective," *International Journal of Sports Physiology and Performance*, 2009, 4, 122–128.
Andrew Kilding et al., "Inspiratory muscle training improves 100 and 200 m swimming performance," *European Journal of Applied Physiology*, 2010, 108, 505–511.
Tom Tong et al., "Chronic and acute inspiratory muscle loading augment the effect of a 6-week interval program on tolerance of high-intensity intermittent bouts of running," *Journal of Strength and Conditioning Research*, 2010, published online ahead of print.

Chapter 3: The Physiology of Exercise
What role does my brain play in fatigue?
Timothy Noakes et al., "From catastrophe to complexity: A novel model of integrative central neural regulation of effort and fatigue during exercise in humans," *British Journal of Sports Medicine*, 2004, 38, 511–514.
Ross Tucker et al., "Impaired exercise performance in the heat is associated with an anticipatory reduction in skeletal muscle recruitment," *Pflügers Archiv—European Journal of Physiology*, 2004, 448(4), 422–430.

Does lactic acid cause muscle fatigue?

A. M. Bellinger et al., "Remodeling of ryanodine receptor complex causes 'leaky' channels: A molecular mechanism for decreased exercise capacity," *PNAS*, 2008, 105(6), 2198–2202.

George Brooks, "Cell-cell and intracellular lactate shuttles," *Journal of Physiology*, 2009, 587(23), 5591–5600.

Why do I get sore a day or two after hard exercise?

Glyn Howatson and Ken van Someren, "The prevention and treatment of exercise-induced muscle damage," *Sports Medicine*, 2008, 38(6), 483–503.

Christer Malm et al., "Leukocytes, cytokines, growth factors and hormones in human skeletal muscle and blood after uphill or downhill running," *Journal of Physiology*, 2004, 556(3), 983–1000.

Shiori Murase et al., "Bradykinin and nerve growth factor play pivotal roles in muscular mechanical hyperalgesia after exercise (delayed-onset muscle soreness)," *Journal of Neuroscience*, 2010, 30(10), 3752–3761.

What is "VO_2max" and should I have mine tested?

Edward Coyle, "Improved muscular efficiency displayed as Tour de France champion matures," *Journal of Applied Physiology*, 2005, 98, 2191–2196.

Benjamin Levine, "VO_2max: What do we know, and what do we still need to know?" *Journal of Physiology*, 2008, 586(1), 25–34.

What is "lactate threshold" and should I have mine tested?

E. P. Debold et al., "Effect of low pH on single skeletal muscle myosin mechanics and kinetics," *American Journal of Physiology—Cell Physiology*, 2008, 295, 173–179.

Oliver Faude et al., "Lactate threshold concepts: How valid are they?" *Sports Medicine*, 2009, 39(6), 469–490.

How can I avoid muscle cramps?

Kevin Miller et al., "Reflex inhibition of electrically induced muscle cramps in hypohydrated humans," *Medicine & Science in Sports & Exercise*, 2010, 42(5), 953–961.

Kevin Miller et al., "Three percent hypohydration does not affect the threshold frequency of electrically-induced cramps," *Medicine & Science in Sports & Exercise*, 2010, published online ahead of print.

Martin Schwellnus, "Cause of exercise associated muscle cramps (EAMC)—altered neuromuscular control, dehydration or electrolyte depletion?" *British Journal of Sports Medicine*, 2009, 43, 401–408.

What's happening when I get a stitch?

Darren Morton, "Exercise related transient abdominal pain," *British Journal of Sports Medicine*, 2003, 37, 287–288.

Darren Morton and Robin Callister, "EMG activity is not elevated during exercise-related transient abdominal pain," *Journal of Science and Medicine in Sport*, 2008, 11, 569–574.

Darren Morton and Robin Callister, "Influence of posture and body type on the experience of exercise-related transient abdominal pain," *Journal of Science and Medicine in Sports*, 2010, 13(5), 485–488.

At what time of day am I strongest and fastest?

Stephen Blonc et al., "Effects of 5 weeks of training at the same time of day on the diurnal variations of maximal muscle power performance," *Journal of Strength and Conditioning Research*, 2010, 24(1), 23–29.

David Hill et al., "Circadian specificity in exercise training," *Ergonomics*, 1989, 32(1), 79–92.

Thomas Reilly and Jim Waterhouse, "Sports performance: Is there evidence that the body clock plays a role?" *European Journal of Applied Physiology*, 2009, 106, 321–332.

Nizar Souissi et al., "Effect of time of day on aerobic contribution to the 30-s Wingate test performance," *Chronobiology International,* 2007, 24(4) 739–748.

Chapter 4: Aerobic Exercise

Why should I do cardio if I just want to build my muscles?

A. G. Monteiro et al., "Acute physiological responses to different circuit training protocols," *Journal of Sports Medicine and Physical Fitness,* 2008, 48(4), 438–442.

Hans-Peter Platzer et al., "Comparison of physical characteristics and performance among elite snowboarders," *Journal of Strength and Conditioning Research,* 2009, 23(5), 1427–1432.

Greg Wells et al., "Physiological correlates of golf performance," *Journal of Strength and Conditioning Research,* 2009, 23(3), 741–750.

How hard should my cardio workout feel?

R. G. Eston et al., "Use of perceived effort ratings to control exercise intensity in young healthy adults," *European Journal of Applied Physiology,* 1987, 56, 222–224.

Carl Foster et al., "The talk test as a marker of exercise training intensity," *Journal of Cardiopulmonary Rehabilitation and Prevention,* 2008, 28(1), 24–30.

How do I determine my maximum heart rate?

Robert Robergs and Roberto Landwehr, "The surprising history of the 'HRmax=22-age' equation," *Journal of Exercise Physiology Online,* 2002, 5(2), 1–10.

Hirofumi Tanaka et al., "Age-predicted maximal heart rate revisited," *Journal of the American College of Cardiology,* 2001, 37(1), 153–156.

Gerald Zavorsky, "Evidence and possible mechanisms of altered maximum heart rate with endurance training and tapering," *Sports Medicine,* 2000, 29(1), 13–26.

What's the best way to breathe during exercise?

R. R. Bechbache and J. Duffin, "The entrainment of breathing frequency by exercise rhythm," *Journal of Physiology*, 1977, 272(3), 553–561.

P. Bernasconi and J. Kohl, "Analysis of co-ordination between breathing and exercise rhythms in man," *Journal of Physiology*, 1993, 471, 693–706.

M. R. Bonsignore et al., "Ventilation and entrainment of breathing during cycling and running in triathletes," *Medicine & Science in Sports & Exercise*, 1998, 30(2), 239–245.

William McDermott et al., "Running training and adaptive strategies of locomotor-respiratory coordination," *European Journal of Applied Physiology*, 2003, 89, 435–444.

Linda Schücker et al., "The effect of attentional focus on running economy," *Journal of Sports Sciences*, 2009, 27(12), 1241–1248.

Will running on hard surfaces increase my risk of injuries?

Daniel Ferris et al., "Running in the real world: Adjusting leg stiffness for different surfaces," *Proceedings of the Royal Society of London B*, 1998, 265, 989–994.

Jack Taunton et al., "A prospective study of running injuries: The Vancouver Sun Run 'In Training' clinics," *British Journal of Sports Medicine*, 2003, 37, 239–244.

Vitor Tessutti et al., "In-shoe plantar pressure distribution during running on natural grass and asphalt in recreational runners," *Journal of Science and Medicine in Sports*, 2010, 13(1), 151–155.

Mark Tillman et al., "In-shoe plantar measurements during running on different surfaces: Changes in temporal and kinetic parameters," *Sports Engineering*, 2002, 5, 121–128.

Do I run "wrong"?

Regan Arendse et al., "Reduced eccentric loading of the knee with the Pose running method," *Medicine & Science in Sports & Exercise*, 2004, 36(2), 272–277.

George Dallam et al., "Effect of a global alteration of running technique on kinematics and economy," *Journal of Sports Sciences*, 2005, 23(7), 757–764.

Bryan Heiderscheit et al., "Effects of step rate manipulation on joint mechanics during running," *Medicine & Science in Sports & Exercise*, 2010, published online before print.

What's the best way to run up and down hills?

William Ebben et al., "Effect of the degree of hill slope on acute downhill running velocity and acceleration," *Journal of Strength and Conditioning Research*, 2008, 22(3), 898–902.

Andrew Townshend et al., "Spontaneous pacing during overground hill running," *Medicine & Science in Sports & Exercise*, 2010, 42(1), 160–169.

Does pumping my arms make me run faster?

Daniel Ferris et al., "Moving the arms to activate the legs," *Exercise and Sport Science Reviews*, 2006, 34(3), 1–8.

Herman Pontzer et al., "Control and function of arm swing in human walking and running," *Journal of Experimental Biology*, 2009, 212, 523–534.

Robert Ropret et al., "Effects of arm and leg loading on sprint performance," *European Journal of Applied Physiology*, 1998, 77, 547–550.

Do spinning classes offer any benefits that I can't get from biking on my own?

Rebecca Battista et al., "Physiologic responses during indoor cycling," *Journal of Strength and Conditioning Research*, 2008, 22(4), 1236–1241.

Will taking the stairs make a real difference to my health?

Colin Boreham et al., "Training effects of short bouts of stair climbing on cardiorespiratory fitness, blood lipids, and homocysteine in sedentary young women," *British Journal of Sports Medicine*, 2005, 39, 590–593.

A. E. Minetti et al., "Skyscraper running: Physiological and biomechanical profile of a novel sports activity," *Scandinavian
Journal of Medicine & Science in Sports*, 2009, published online
ahead of print.

Chapter 5: Strength and Power

Do I need strength training if I just want to be lean and fit?

Stuart Phillips, "Resistance exercise: Good for more than just
Grandma and Grandpa's muscles," *Applied Physiology,
Nutrition, and Metabolism*, 2007, 32, 1198–1205.

How much weight should I lift, and how many times?

Sofia Bågenhammar and Eva Jansson, "Repeated sets or single
set of resistance training—A systematic review," *Advances in
Physiotherapy*, 2007, 9(4), 154–160.

Nicholas Burd et al., "Low-load high volume resistance exercise
stimulates muscle protein synthesis more than high-load low
volume resistance exercise in young men," *PLoS ONE*, 2010,
5(8), e12033.

William Kraemer et al., "Progression models in resistance training
for healthy adults," American College of Sports Medicine position stand, 2002.

Sharon Rana et al., "Comparison of early phase adaptations for
traditional strength and endurance, and low velocity resistance
training programs in college-aged women," *Journal of Strength
and Conditioning Research*, 2008, 22(1), 119–127.

How do I tone my muscles without bulking up?

Stephen Glass and Douglas Stanton, "Self-selected resistance
training intensity in novice weightlifters," *Journal of Strength
and Conditioning Research*, 2004, 18(2), 324–327.

Nicholas Ratamess et al., "Self-selected resistance training intensity in healthy women: The influence of a personal trainer,"
Journal of Strength and Conditioning Research, 2008, 22(1),
103–111.

What's the difference between strength and power?
Greg Wells et al., "Physiological correlates of golf perform-
    ance," *Journal of Strength and Conditioning Research,* 2009, 23(3),
    741–750.
David Szymanski et al., "Contributing factors for increased bat
    swing velocity," *Journal of Strength and Conditioning Research,*
    2009, 23(4), 1338–1352.

**Free weights or machines: what's the difference, and which
should I use?**
Michael Wahl and David Behm, "Not all instability training de-
    vices enhance muscle activation in highly resistance-trained
    individuals," *Journal of Strength and Conditioning Research,* 2008,
    22(4), 1360–1370.

**Can body-weight exercises like push-ups and sit-ups be as
effective as lifting weights?**
N. L. Ashworth et al., "Home versus center based physical activity
    programs in older adults," Cochrane Library, 2009.
Junichiro Yamauchi et al., "Effects of bodyweight-based exercise
    training on muscle functions of leg multi-joint movement in
    elderly individuals," *Geriatrics and Gerontology,* 2009, 9, 262–269.
James Youdas et al., "Comparison of muscle-activation patterns
    during conventional push-up and Perfect Pushup exercises,"
    Journal of Strength and Conditioning Research, 2010, published
    online ahead of print.

Can lifting weights fix my lower-back pain?
Stanley Bigos et al., "High-quality controlled trials on preventing
    episodes of back problems: Systematic literature review in
    working-age adults," *Spine Journal,* 2009, 9, 147–168.
J. Hayden et al., "Exercise therapy for treatment of non-specific
    low back pain," Cochrane Library, 2010.
Joel Jackson et al., "The influence of periodized resistance train-
    ing on recreationally active males with chronic nonspecific low

back pain," *Journal of Strength and Conditioning Research*, 2010, published online ahead of print.

Robert Kell et al., "The response of persons with chronic nonspecific low back pain to 3 different volumes of periodized musculoskeletal rehabilitation," *Journal of Strength and Conditioning Research*, 2010, published online ahead of print.

Will I get a better workout if I hire a personal trainer?

Paulo Gentil and Martim Bottaro, "Influence of supervision ratio on muscle adaptations to resistance training in nontrained subjects," *Journal of Strength and Conditioning Research*, 2010, 24(3), 639–643.

Scott Mazzetti et al., "The influence of direct supervision of resistance training on strength performance," *Medicine & Science in Sports & Exercise*, 2000, 32(6), 1175–1184.

Nicholas Ratamess et al., "Self-selected resistance training intensity in healthy women: The influence of a personal trainer," *Journal of Strength and Conditioning Research*, 2008, 22(1), 103 111.

Do I need extra protein to build muscle?

D. R. Moore et al., "Ingested protein dose response of muscle and albumin protein synthesis after resistance exercise in young men," *American Journal of Clinical Nutrition*, 2009, 89(1), 161–168.

Stuart Phillips et al., "A critical examination of dietary protein requirements, benefits, and excesses in athletes," *International Journal of Sport Nutrition and Exercise Metabolism*, 2007, 17(4S).

Mark Tarnopolsky, "Protein requirements for endurance athletes," *Nutrition*, 2004, 20(7-8), 662–668.

Chapter 6: Flexibility and Core Strength

Will stretching help me avoid injuries?

Ian Shrier, "Stretching before exercise: An evidence based approach," *British Journal of Sports Medicine*, 2000, 34, 324–325.

Katie Small et al., "A systematic review into the efficacy of static stretching as part of a warm-up for the prevention of exercise-related injury," *Research in Sports Medicine,* 2008, 16(3), 213–231.

Stephen Thacker et al., "The impact of stretching on sports injury risk: A systematic review of the literature," *Medicine & Science in Sports & Exercise,* 2004, 36(3), 371–378.

Could stretching before exercise make me slower and weaker?

J. R. Fowles et al., "Reduced strength after passive stretch of the human plantarflexors," *Journal of Applied Physiology,* 2000, 89(3), 1179–1188.

Antonio La Torre et al., "Acute effects of static stretching on squat jump performance at different knee starting angles," *Journal of Strength and Conditioning Research,* 2010, 24(3), 687–694.

P. C. Macpherson et al., "Contraction-induced injury to single fiber segments from fast and slow muscles of rats by single stretches," *American Journal of Physiology,* 271, C1438–1446.

Jason Winchester et al., "Static stretching impairs sprint performance in collegiate track and field athletes," *Journal of Strength and Conditioning Research,* 2008, 22(1), 13–19.

Do flexible runners run more efficiently?

A. M. Jones, "Running economy is negatively related to sit-and-reach test performance in international-standard distance runners," *International Journal of Sports Medicine,* 2002, 23, 40–43.

Tamra Trehearn et al., "Sit-and-reach flexibility and running economy of men and women collegiate distance runners," *Journal of Strength and Conditioning Research,* 2009, 23(1), 158–162.

Jacob Wilson et al., "Effects of static stretching on energy cost and running endurance performance," *Journal of Strength and Conditioning Research,* 2009, published online ahead of print.

How should I warm up before exercise?

Sonja Herman and Derek Smith, "Four-week dynamic stretching warm-up intervention elicits longer-term performance

benefits," *Journal of Strength and Conditioning Research,* 22(4),
1286–1297.

Danny McMillian et al., "Dynamic vs. static-stretching warm
up: The effect on power and agility performance," *Journal of
Strength and Conditioning Research,* 2006, 20(3), 492–499.

Will stretching after exercise help me avoid next-day soreness?

Maarten Bobbert et al., "Factors in delayed onset muscular soreness of man," *Medicine & Science in Sports & Exercise,* 1986,
18(1), 75–81.

B. Dawson et al., "Effects of immediate post-game recovery procedures on muscle soreness, power and flexibility levels over
the next 48 hours," *Journal of Science and Medicine in Sport,* 2005,
8(2), 210–221.

R. D. Herbert and M. de Noronha, "Stretching to prevent or
reduce muscle soreness after exercise," The Cochrane Library,
2008.

Elisa Robey et al., "Effect of postexercise recovery procedures
following strenuous stairclimb running," *Research in Sports
Medicine,* 2009, 17(4), 245–259.

Where is my "core," and do I need to strengthen it?

Darin Leetun et al., "Core stability measures as risk factors for
lower extremity injury in athletes," *Medicine & Science in Sports
& Exercise,* 2004, 36(6), 926–934.

Michele Olson et al., "Prediction of superficial versus deep abdominal muscle activity during selected Pilates exercises,"
Medicine & Science in Sports & Exercise, 2008, 40(5), S426.

What are the benefits of yoga for physical fitness?

B. Donohue et al., "Effects of brief yoga exercises and motivational
preparatory interventions in distance runners: Results of a controlled trial," *British Journal of Sports Medicine,* 2006, 40, 60–63.

Marshall Hagins et al., "Does practicing yoga satisfy recommendations for intensity of physical activity which improves

and maintains health and cardiovascular fitness?" *BMC Complementary and Alternative Medicine*, 2007, 7(40).

Meg Hayes and Sam Chase, "Prescribing yoga," *Primary Care: Clinics in Office Practice*, 2010, 37, 31–47.

Mark Tran et al., "Effects of hatha yoga practice on the health-related aspects of physical fitness," *Preventive Cardiology*, 2001, 4, 165–170.

What are the benefits of yoga for overall wellness?

P. R. Bosch et al., "Functional and physiological effects of yoga in women with rheumatoid arthritis: A pilot study," *Alternative Therapies in Health and Medicine*, 2009, 15(4), 24–21.

Joseph Pellegrino et al., "Differential affective responses to acute hatha yoga and moderate intensity resistance training," *Medicine & Science in Sports & Exercise*, 2008, 40(5), S17.

Alyson Ross and Sue Thomas, "The health benefits of yoga and exercise: A review of comparison studies," *Journal of Alternative and Complementary Medicine*, 2010, 16(1), 3–12.

H. S. Vadiraja et al., "Effects of a yoga program on cortisol rhythm and mood states in early breast cancer patients undergoing adjuvant radiotherapy: A randomized controlled trial," *Integrative Cancer Therapies*, 2009, 8(1), 37–46.

Chapter 7: Injuries and Recovery

Ouch, I think I sprained something. How long should I stay off it?

F. E. Faria et al., "The onset and duration of mobilization affect the regeneration in the rat muscle," *Histology and Histopathology*, 2008, 23(5), 565–571.

Tero Järvinen et al., "Muscle injuries: Optimising recovery," *Best Practice & Research Clinical Rheumatology*, 2007, 21(2), 317–331.

P. Kannus et al., "Basic science and clinical studies coincide: Active treatment approach is needed after a sports injury," *Scandinavian Journal of Medicine and Science in Sports*, 2003, 13, 150–154.

Michael Wolfe et al., "Management of ankle sprains," *American Family Physician*, 2001, 63(1), 93–105.

Will a post-exercise ice bath help me recover more quickly?
Jeremy Ingram et al., "Effect of water immersion methods on post-exercise recovery from simulated team sport exercise," *Journal of Science and Medicine in Sport*, 2009, 12, 417–421.
Greg Rowsell et al., "Effects of cold-water immersion on physical performance between successive matches in high-performance junior male soccer players," *Journal of Sports Sciences*, 2009, 27(6), 565–573.
Kylie Sellwood et al., "Ice-water immersion and delayed-onset muscle soreness: A randomized controlled trial," *British Journal of Sports Medicine*, 2007, 41, 392–397.

Will a heat pack or hot bath soothe my aching body?
S. D. French et al., "Superficial heat or cold for low back pain," Cochrane Library, 2010.
Kei Mizuno et al., "Effects of mild-stream bathing on recovery from mental fatigue," *Medical Science Monitor*, 2010, 16(1), CR8–14.
Val Robertson et al., "The effect of heat of tissue extensibility: A comparison of deep and superficial heating," *Archives of Physical Medicine and Rehabilitation*, 2005, 86, 819–825.

Will massage help me avoid soreness and recover more quickly from workouts?
P. Bakowski et al., "Effects of massage on delayed-onset muscle soreness," *Chir Narzadow Ruchu Ortop Pol*, 2008, 73(4), 261–265.
T. A. Butterfield et al., "Cyclic compressive loading facilitates re-covery after eccentric exercise," *Medicine & Science in Sports & Exercise*, 2008, 40(7), 1289–1296.
E. V. Wiltshire et al., "Massage impairs postexercise muscle blood flow and 'lactic acid' removal," *Medicine & Science in Sports & Exercise*, 2010, 42(6), 1062–1071.

Should I take painkillers for post-workout soreness?
Tatiane Gorski et al., "Use of nonsteroidal anti-inflammatory drugs (NSAIDs) in triathletes: Prevalence, level of awareness, and reasons for use," *British Journal of Sports Medicine*, 2009, published online ahead of print.
David Nieman et al., "Ibuprofen use, endotoxemia, inflammation, and plasma cytokines during ultramarathon competition," *Brain, Behavior, and Immunity*, 2006, 20, 578–584.
S. P. Tokmakidis et al., "The effects of ibuprofen on delayed muscle soreness and muscular performance after eccentric exercise," *Journal of Strength and Conditioning Research*, 2003, 17(1), 53–59.
Stuart Warden, "Prophylactic misuse and recommended use of non-steroidal anti-inflammatory drugs by athletes," *British Journal of Sports Medicine*, 2009, 43, 548–549.

How long does it take to recover after a marathon or other long, intense effort?
N. Mousavi et al., "Relation of biomarkers and cardiac magnetic resonance imaging after marathon running," *American Journal of Cardiology*, 2009, 103(10), 1467–1472.
Kim Petersen, "Muscle mechanical characteristics in fatigue and recovery from a marathon race in highly trained runners," *European Journal of Applied Physiology*, 2007, 101, 385–396.
William Sherman et al., "Effect of a 42.2-km footrace and subsequent rest or exercise on muscular strength and work capacity," *Journal of Applied Physiology: Respiratory, Environmental and Exercise Physiology*, 1984, 57(6), 1668–1673.

Can "platelet-rich plasma" cure my tennis elbow or Achilles tendon?
Robert de Vos et al., "Platelet-rich plasma injection for chronic Achilles tendinopathy: A randomized controlled trial," *JAMA*, 2010, 303(2), 144–149.
Allan Mishra et al., "Treatment of tendon and muscle using platelet-rich plasma," *Clinics in Sports Medicine*, 2009, 28, 113–125.

Joost Peerbooms et al., "Positive effect of an autologous platelet concentrate in lateral epicondylitis in a double-blind randomized controlled trial: Platelet-rich plasma versus corticosteroid injection with a 1-year follow-up," *American Journal of Sports Medicine,* 2010, 38, 255–262.

How can I reduce my risk of stress fractures?

Brent Edwards et al., "Effects of stride length and running mileage on a probabilistic stress fracture model," *Medicine & Science in Sports & Exercise,* 2009, 41(12), 2177–2184.

Kristin Popp et al., "Bone geometry, strength, and muscle size in runners with a history of stress fracture," *Medicine & Science in Sports & Exercise,* 2009, 41(12), 2145–2150.

Should I exercise when I'm sick?

Thomas Weidner et al., "Effect of a rhinovirus-caused upper respiratory illness on pulmonary function test and exercise responses," *Medicine & Science in Sports & Exercise,* 1997, 29(5), 604–609.

Thomas Weidner et al., "The effect of exercise training on the severity and duration of a viral upper respiratory illness," *Medicine & Science in Sports & Exercise,* 1998, 30(11), 1578–1583.

Will having a few drinks affect my workout the next day?

Matthew Barnes et al., "Acute alcohol consumption aggravates the decline in muscle performance following strenuous eccentric exercise," *Journal of Science and Medicine in Sport,* 2010, 13, 189–193.

Michael Stein and Peter Friedmann, "Disturbed sleep and its relationship to alcohol use," *Substance Abuse,* 2005, 26(1), 1–13.

Chapter 8: Exercise and Aging

What's the cumulative effect of all the exercise I've done over the years?

Wojtek Chodzko-Zajko et al., "Exercise and physical activity for older adults," American College of Sports Medicine position stand, 2009.

M. H. Elleuch et al., "Knee osteoarthritis in 50 former top-level soccer players: A comparative study," *Annales de Réadaptation et de Médecine Physique,* 2008, 51(3), 174–178.

N. Thelin et al., "Knee injuries account for the sports-related increased risk of knee osteoarthritis," *Scandinavian Journal of Medicine and Science in Sports,* 2006, 16(5), 329–333.

Paul Williams, "Relationship of incident glaucoma versus physical activity and fitness in male runners," *Medicine & Science in Sports & Exercise,* 2009, 41(8), 1566–1572.

Will running ruin my knees?

Eliza Chakravarty et al., "Long distance running and knee osteoarthritis: A prospective study," *American Journal of Preventive Medicine,* 2008, 35(2), 133–138.

David Felson et al., "Effect of recreational physical activities on the development of knee osteoarthritis in older adults of different weights: The Framingham Study," *Arthritis & Rheumatism,* 2007, 57(1), 6–12.

Wolfgang Krampla et al., "Changes on magnetic resonance tomography in the knee joints of marathon runners: A 10-year longitudinal study," *Skeletal Radiology,* 2008, 37(7) 619–626.

How should I adapt my workout routine as I get older?

Carl Foster et al., "Training in the aging athlete," *Current Sports Medicine Reports,* 2007, 6, 200–206.

Bradley Young et al., "Explaining performance in elite middle-aged runners: Contributions from age and from ongoing and past training factors," *Journal of Sports and Exercise Psychology,* 2008, 30(6), 737–754.

How quickly will my performance decline as I age?

Maria Anton et al., "Age-related declines in anaerobic muscular performance: Weightlifting and powerlifting," *Medicine & Science in Sports & Exercise,* 2004, 36(1), 143–147.

M. J. Stones and A. Kozma, "Cross-sectional, longitudinal, and secular age trends in athletic performance," *Experimental Aging Research*, 1982, 8, 185–188.

Hirofumi Tanaka and Douglas Seals, "Endurance exercise performance in Masters athletes: Age-associated changes and underlying physiological mechanisms," *Journal of Physiology*, 2008, 586(1), 55–63.

Bradley Young et al., "Does lifelong training temper age-related decline in sport performance? Interpreting differences between cross-sectional and longitudinal data," *Experimental Aging Research*, 2008, 34(1), 27–48.

How can I stay motivated to exercise as my performances decline?

N. A. Christakis and J. H. Fowler, "The spread of obesity in a large social network over 32 years," *New England Journal of Medicine*, 2007, 357(4), 370–379.

Bradley Young, "Psycho-social perspectives on commitment, maintenance and performance in masters sport," Master and Mentors conference, Lahti, Finland, 2009.

What are the pros and cons of exercising in water?

E. M. Bartels et al., "Aquatic exercise for the treatment of knee and hip osteoarthritis," Cochrane Library, 2009.

Danilo Bocalini et al., "Water- versus land-based exercise effects on physical fitness in older women," *Geriatrics and Gerontology International*, 2008, 8, 265–271.

Umit Dundar et al., "Clinical effectiveness of aquatic exercise to treat chronic low back pain," *Spine*, 2009, 34(14), 1436–1440.

Laurent Mourot et al., "Training-induced increase in nitric oxide metabolites in chronic heart failure and coronary artery disease: An extra benefit of water-based exercise?" *European Journal of Cardiovascular Prevention and Rehabilitation*, 2009, 16, 215–221.

What type of exercise is best for maintaining strong bones?

H. M. Macdonald et al., "Is a school-based physical activity inter-vention effective for increasing tibial bone strength in boys and girls?" *Journal of Bone and Mineral Research*, 2007, 22(3), 434–446.

Scott Rector et al., "Lean body mass and weight-bearing activity in the prediction of bone mineral density in physically active men," *Journal of Strength and Conditioning Research*, 2009, 23(2), 427–435.

Can exercise keep my DNA from aging?

Ramin Farzaneh-Far et al., "Prognostic value of leukocyte telo-mere length in patients with stable coronary artery disease," *Arteriosclerosis, Thrombosis, and Vascular Biology*, 2008, 28, 1379–1384.

Thomas LaRocca et al., "Leukocyte telomere length is preserved with aging in endurance exercise-trained adults and related to maximal aerobic capacity," *Mechanisms of Ageing and Development*, 2010, 131, 165–167.

Christian Werner et al., "Physical exercise prevents cellular senescence in circulating leukocytes and in the vessel wall," *Circulation*, 2009, 120, 2438–2447.

Chapter 9: Weight Management

Is it possible to be fat and healthy at the same time?

Heather Orpana et al., "BMI and mortality: Results from a national longitudinal study of Canadian adults," *Obesity*, 2010, 18(1), 214–218.

C. D. Lee et al., "Cardiorespiratory fitness, body composition, and all-cause mortality in men," *American Journal of Clinical Nutrition*, 1999, 69(3), 373–380.

Is weight loss simply the difference between "calories in" and "calories out"?

Rochelle Goldsmith et al., "Effects of experimental weight per-turbation on skeletal muscle work efficiency, fuel utilization, and biochemistry in human subjects," *American Journal of*

Physiology—Regulatory, Integrative and Comparative Physiology, 2010, 298, R79–R88.

Martijn Katan and David Ludwig, "Extra calories cause weight gain but how much?" *JAMA,* 2010, 303(1), 65–66.

To lose weight, is it better to eat less or exercise more?

Enette Larson-Meyer et al., "Caloric restriction with or without exercise: The fitness versus fatness debate," *Medicine & Science in Sports & Exercise,* 2010, 42(1), 152–159.

How can I take advantage of the "fat-burning" zone?

Kyle Hoehn et al., "Acute or chronic upregulation of mitochondrial fatty acid oxidation has no net effect on whole-body energy expenditure or adiposity," *Cell Metabolism,* 2010, 11, 70–76.

Asker Jeukendrup et al., "Fat burning: How and why?" in *Sports Nutrition: From Lab to Kitchen,* Meyer & Meyer Sport, 2010.

Won't exercise make me eat more and gain weight?

John Cloud, "Why exercise won't make you thin," *Time,* August 9, 2009.

I-Min Lee et al., "Physical activity and weight gain prevention," *JAMA,* 2010, 303(12), 1173–1179.

Can I lose weight while gaining (or maintaining) muscle?

Donald Layman et al., "Dietary protein and exercise have additive effects on body composition during weight loss in adult women," *Journal of Nutrition,* 2005, 135(8), 1903–1910.

P. W. Macdermid and S. R. Stannard, "A whey-supplemented, high-protein diet versus a high-carbohydrate diet: Effects on endurance cycling performance," *International Journal of Sports Nutrition and Exercise Metabolism,* 2006, 16(1), 65–77.

Samuel Mettler et al., "Increased protein intake reduces lean body mass during weight loss in athletes," *Medicine & Science in Sports & Exercise,* 2010, 42(2), 326–337.

Is lifting weights better than cardio for weight loss?

Joseph Donnelly et al., "Appropriate physical activity intervention strategies for weight loss and prevention of weight regain for adults," American College of Sports Medicine position stand, 2009.

Kathyrn Schmitz et al., "Strength training and adiposity in pre-menopausal women: Strong, Healthy, and Empowered study," *American Journal of Clinical Nutrition*, 2007, 86, 566–572.

S. K. Park et al., "The effect of combined aerobic and resistance exercise training on abdominal fat in obese middle-aged women," *Journal of Physiological Anthropology and Applied Human Science*, 2003, 22(3), 129–135.

S. P. Tzankoff and A. H. Norris, "Effect of muscle mass decrease on age-related BMR changes," *Journal of Applied Physiology: Respiratory, Environmental and Exercise Physiology*, 1977, 43(6), 1001–1006.

Will I burn more calories commuting by bike or on foot?

C. Hall et al., "Energy expenditure of walking and running: Comparison with prediction equations," *Medicine & Science in Sports & Exercise*, 2004, 36(12), 2128–2134.

Karen Steudel-Numbers and Cara Wall-Scheffler, "Optimal running speed and the evolution of hominin hunting strategies," *Journal of Human Evolution*, 2009, 56(4), 355–360.

David Swain, "The influence of body mass in endurance bicycling," *Medicine & Science in Sports & Exercise*, 1994, 26(1), 58–63.

Darren Warburton et al., "Prescribing exercise as preventive therapy," *Canadian Medical Association Journal*, 2006, 174(7), 961–974.

Can I control hunger by manipulating my appetite hormones?

Jameason Cameron et al., "Increased meal frequency does not promote greater weight loss in subjects who were prescribed an 8-week equi-energetic energy-restricted diet," *British Journal of Nutrition*, 2010, 103(8), 1098–1101.

Alexander Kokkinos et al., "Eating slowly increases the post-prandial response of the anorexigenic gut hormones, peptide YY and glucagon-like peptide-1," *Journal of Clinical Endocrinology and Metabolism*, 2010, 95(1), 333–337.

Will sitting too long at work counteract all my fitness gains?
E. A. Beers et al., "Increasing passive energy expenditure during clerical work," *European Journal of Applied Physiology*, 2008, 103 (3), 353–360.
R. E. Macpherson et al., "Run sprint interval training improves aerobic performance but not max cardiac output," *Medicine & Science in Sports & Exercise*, 2010, published online ahead of print.
Alpa Patel et al., "Leisure time spent sitting in relation to total mortality in a prospective cohort of US adults," *American Journal of Epidemiology*, 2010, 172 (4), 419–429.
Gretchen Reynolds, "Weighing the evidence on exercise," *New York Times*, April 16, 2010.

Chapter 10: Nutrition and Hydration

Should I carbo-load by eating pasta the night before a competition?

Louise Burke et al., "Nutrition for distance events," *Journal of Sports Sciences*, 2007, 25(1), S29–S38.
Mark Hargreaves et al., "Pre-exercise carbohydrate and fat ingestion: Effects on metabolism and performance," *Journal of Sports Sciences*, 2004, 22(1), 31–38.
Asker Jeukendrup, "The optimal pre-competition meal," in *Sports Nutrition: From Lab to Kitchen*, Meyer & Meyer Sport, 2010.

What should I eat to avoid stomach problems during exercise?

Alon Harris et al., "Rapid orocecal transit in chronically active persons with high energy intake," *Journal of Applied Physiology*, 1991, 70(4), 1550–1553.
Robert Murray, "Training the gut for competition," *Current Sports Medicine Reports*, 2006, 5, 161–164.

Beate Pfeiffer, "Nutrition- and exercise-associated gastrointestinal problems," in *Sports Nutrition: From Lab to Kitchen*, Meyer & Meyer Sport, 2010.

What should I eat and drink to refuel after working out?
Steven Black et al., "Improved insulin action following short-term exercise training: Role of energy and carbohydrate balance," *Journal of Applied Physiology*, 2005, 99, 2285–2293.
Louise Burke, "Nutrition for recovery," in *Sports Nutrition: From Lab to Kitchen*, Meyer & Meyer Sport, 2010.

How much should I drink to avoid dehydration during exercise?
D. Passe et al., "Voluntary dehydration in runners despite favorable conditions for fluid intake," *International Journal of Sports Nutrition and Exercise Metabolism*, 2007, 17(3), 284–295.
Michael Sawka and Timothy Noakes, "Does dehydration impair exercise performance?" *Medicine & Science in Sports & Exercise*, 2007, 39(8), 1209–1217.

Is it possible to hydrate too much?
Courtney Kipps et al., "The incidence of exercise-associated hyponatraemia in the London Marathon," *British Journal of Sports Medicine*, 2010, published online ahead of print.
T. D. Noakes and D. B. Speedy, "Case proven: Exercise associated hyponatraemia is due to overdrinking. So why did it take 20 years before the original evidence was accepted?" *British Journal of Sports Medicine*, 2006, 40, 567–572.

What ingredients do I really need in a sports drink?
E. S. Chambers et al., "Carbohydrate sensing in the human mouth: Effects on exercise performance and brain activity," *Journal of Physiology*, 2009, 587(8), 1779–1794.
A. E. Jeukendrup, "Effect of beverage glucose and sodium content on fluid delivery," *Nutrition and Metabolism*, 2009, 20(6), 9.

Will taking antioxidant vitamins block the health benefits of exercise?

Michael Ristow et al., "Antioxidants prevent health-promoting effects of physical exercise in humans," *PNAS*, 2009, 106(21), 8665–8670.

Vitor Teixeira et al., "Antioxidants do not prevent postexercise peroxidation and may delay muscle recovery," *Medicine & Science in Sports & Exercise*, 2009, 41(9), 1752–1760.

A. E. Carrillo et al., "Vitamin C supplementation and salivary immune function following exercise-heat stress," *International Journal of Sports Physiology and Performance*, 2008, 3(4), 516–530.

Should I be taking probiotics?

A. J. Cox et al., "Oral administration of the probiotic *Lactobacillus fermentum* VRI-003 and mucosal immunity in endurance athletes," 2010, 44, 222–226.

R. A. Kekkonen et al., "The effect of probiotics on respiratory infections and gastrointestinal symptoms during training in marathon runners," *International Journal of Sport Nutrition and Exercise Metabolism*, 2007, 17, 352–363.

N. P. West et al., "Probiotics, immunity and exercise: A review," *Exercise Immunology Review*, 2009, 15, 107–126.

Will vitamin D make me a better athlete?

John Cannell et al., "Athletic performance and vitamin D," *Medicine & Science in Sports & Exercise*, 2009, 41(5), 1102–1110.

Anne Looker et al., "Serum 25-hydroxyvitamin D status of the US population: 1988–1994 compared with 2000–2004," *American Journal of Clinical Nutrition*, 2008, 88, 1519–1527.

Kate Ward et al., "Vitamin D status and muscle function in postmenarchal adolescent girls," *Journal of Clinical Endocrinology and Metabolism*, 2009, 94(2), 559–563.

Is there any benefit to deliberately training with low energy stores?

Louise Burke, "New issues in training and nutrition: train low, compete high?" *Current Sports Medicine Reports*, 2007, 6, 137–138.

Anne Hansen et al., "Skeletal muscle adaptation: Training twice every second day vs. training once daily," *Journal of Applied Physiology*, 2005, 98, 93–99.

Carl Hulston et al., "Training with low muscle glycogen enhances fat metabolism in well-trained cyclists," *Medicine & Science in Sports & Exercise*, 2010, published online ahead of print.

Can I get the nutrients I need for a heavy exercise regimen from a vegetarian or vegan diet?

Joel Fuhrman and Deana Ferreri, "Fueling the vegetarian (vegan) athletes," *Current Sports Medicine Reports*, 2010, 9 (4), 233–241.

Chapter 11: Mind and Body

If my brain is tired, will my body's performance suffer?

Samuele Marcora et al., "Mental fatigue impairs physical performance in humans," *Journal of Applied Physiology*, 2009, 106, 857–864.

Does it matter what I'm thinking about when I train?

Anders Ericsson et al., "The role of deliberate practice in the acquisition of expert performance," *Psychological Review*, 1993, 100(3), 363–406.

B. W. Young and J. H. Salmela, "Perceptions of training and deliberate practice of middle distance runners," *International Journal of Sport Psychology*, 2002, 33, 167–181.

Does listening to music or watching TV help or hurt my workout?

H. B. T. Lim, "Effects of differentiated music on cycling time trial," *International Journal of Sports Medicine*, 2009, 30, 435–442.

V. M. Nethery, "Competition between internal and external
    sources of information during exercise: Influence on RPE and
    the impact of the exercise load," *Journal of Sports Medicine and
    Physical Fitness*, 2002, 42(2), 172–178.
J. M. Waterhouse et al., "Effects of music tempo upon submaximal
    cycling performance," *Scandinavian Journal of Medicine and
    Science in Sports*, 2010, 20(4), 662–669.

Will I perform better under pressure if I focus harder?

Sian Beilock et al., "When paying attention becomes counter-
    productive: Impact of divided versus skill-focused attention on
    novice and experienced performance of sensorimotor skills,"
    Journal of Experimental Biology: Applied, 2002, 8(1), 6–16.
Sian Beilock et al., "When does haste make waste? Speed-accuracy
    tradeoff, skill level, and the tools of the trade," *Journal of
    Experimental Biology: Applied*, 2008, 14(4), 340–352.
Linda Schücker et al., "The effect of attentional focus on running
    economy," *Journal of Sports Sciences*, 2009, 27(12), 1241–1248.

Can swearing help me push harder in a workout?

Richard Stephens et al., "Swearing as a response to pain,"
    NeuroReport, 2009, 20, 1056–1060.
Kurt Gray, "Moral transformation: Good and evil turn the weak
    into the mighty," *Social Psychological & Personality Science*, 2010,
    1(3), 253–258.

Is there such a thing as "runner's high"?

Robin Kanarek et al., "Running and addiction: Precipitated with-
    drawal in a rat model of activity-based anorexia," *Behavioral
    Neuroscience*, 2009, 123(4), 905–912.
Henning Boecker et al., "The runner's high: Opioidergic mechan-
    isms in the human brain," *Cerebral Cortex*, 2008, 18, 2523–2531.

Will taking a fitness class or joining a team change my brain chemistry during workouts?

Emma Cohen et al., "Rowers' high: Behavioural synchrony is correlated with elevated pain thresholds," *Biology Letters*, 2010, 6(1), 106–108.

S. N. Fraser and K. S. Spink, "Examining the role of social support and group cohesion in exercise compliance," *Journal of Behavioral Medicine*, 2002, 36(4), 305–312.

What are the effects of exercise on the brain?

Maria Åberg et al., "Cardiovascular fitness is associated with cognition in young adulthood," *PNAS*, 2009, 106(49), 20906–20911.

E. Bullitt et al., "The effect of exercise on the cerebral vasculature of healthy aged subjects as visualized by MR angiography," *American Journal of Neuroradiology*, 2009, 30, 1857–1863.

Teal Eich and Janet Metcalfe, "Effects of the stress of marathon running on implicit and explicit memory," *Psychonomic Bulletin and Review*, 2009, 16(3), 475–479.

Matthew Pontifex et al., "The effect of acute aerobic and resistance exercise on working memory," *Medicine & Science in Sports & Exercise*, 41(4), 927–934.

Yu-Fan Liu et al., "Differential effects of treadmill running and wheel running on spatial or aversive learning and memory: Roles of amygdalar brain-derived neurotrophic factor and synaptotagmin I," *Journal of Physiology*, 2009, 587, 3221–3231.

Chapter 12: The Competitive Edge

How should I adjust my training in the final days before a competition?

Laurent Bosquet et al., "Effects of tapering on performance: A meta-analysis," *Medicine & Science in Sports & Exercise*, 2007, 39(8), 1358–1365.

J. Ekstrand et al., "A congested football calendar and the well-being of players: Correlation between match exposure of European footballers before the World Cup 2002 and their injuries and performances during that World Cup," *British Journal of Sports Medicine,* 2004, 38, 493–497.

David Pyne et al., "Peaking for optimal performance: Research limitations and future directions," *Journal of Sports Sciences,* 2009, 27(3), 195–202.

Should I have sex the night before a competition?

S. McGlone and I. Shrier, "Does sex the night before the competition decrease performance?" *Clinical Journal of Sports Medicine,* 2000, 10(4), 233–234.

Can drinking slushies boost my performance on hot days?

Rodney Siegel et al., "Ice slurry ingestion increases core temperature capacity and running time in the heat," *Medicine & Science in Sports & Exercise,* 42(4), 717–725.

Will drinking coffee help or hinder my performance?

N. Chester and N. Wojek, "Caffeine consumption amongst British athletes following changes to the 2004 WADA prohibited list," *International Journal of Sports Medicine,* 2008, 29, 524–528.

Mark Stuart, "Do athletes and WADA differ in their perception of caffeine as a performance-enhancing drug?" *BMJ Clinical Evidence,* July 28, 2008.

Does competing in front of a crowd help improve performance?

Niels van de Ven, "Supporters are not necessary for the home advantage: Evidence from same-stadium derbies and games without an audience," *Journal of Applied Social Psychology,* 2010, published online ahead of print.

How much sleep do I need for optimal physical performance?

G. S. Passos et al., "Effect of acute physical exercise on patients with chronic primary insomnia," *Journal of Clinical Sleep Medicine,* 2010, 15(6), 270–275.

S. D. Youngstedt, "Effects of exercise on sleep," *Clinics in Sports Medicine*, 2005, 24(2), 355–365.

How should I pace myself in a long-distance race?
A. E. Gosztyla et al., "The impact of different pacing strategies on five-kilometer running time trial performance," *Journal of Strength and Conditioning*, 2006, 20(4), 882–886.
W. J. Rejeski and P. M. Ribisl, "Expected task duration and perceived effort: An attributional analysis," *Journal of Sport Psychology*, 1980, 2, 227–236.
Ross Tucker et al., "An analysis of pacing strategies during men's world-record performances in track athletics," *International Journal of Sports Physiology and Performance*, 2006, 1, 233–245.

Is endurance or sprint speed more important in field sports like soccer?
Martin Buchheit et al., "Improving repeated sprint ability in young elite soccer players: Repeated shuttle sprints vs. explosive strength training," *Journal of Strength and Conditioning Research*, 2010, published online ahead of print.
Stuart Phillips, "Team sports," in *Sports Nutrition: From Lab to Kitchen*, Meyer & Meyer Sport, 2010.
Matt Spencer et al., "Physiological and metabolic responses of repeated-sprint activities specific to field-based team sports," *Sports Medicine*, 2005, 35(12), 1025–1044.

Illustration references

Setting the treadmill incline (p. 33): Jones et al., *Journal of Sports Sciences*, 1996.
The calf muscle pump (p. 43): Bringard et al., *Science & Sports*, 2007.
Adding balance training to your workout (p. 49): Con Hrysomallis, Victoria University.
VO_2max and lactate threshold (p. 65): Tim Noakes, "The Lore of Running"; Martin and Coe, "Better Training for Distance Runners."

Adapting to different running surfaces (p. 87): Ferris et al., Proceedings of the Royal Society B, 1998.

Building power (p. 107): Greg Wells, "Physical Preparation for Golf."

Stability for strength training (p. 110): Wahl et al., *Journal of Strength and Conditioning Research*, 2008.

Dynamic stretching (p. 130): McMillian et al., *Journal of Strength and Conditioning Research*, 2006.

Hip strengthening (p. 135): Reed Ferber, University of Calgary.

Active rehabilitation: Sprained Ankle (p. 144): Wolfe et al., *American Family Physician*, 2001.

The aging body (p. 166): Chodzko-Zajko et al., *Medicine & Science in Sports & Exercise*, 2009.

Slowing down (p. 173): World Masters Athletics

Bone health hot spots (p. 179): Heather McKay, University of British Columbia.

Commuting calories (p. 200): Warburton et al., *Canadian Medical Association Journal*, 2006.

Index

© Lauren King

ALEX HUTCHINSON is a contributing editor at *Popular Mechanics* magazine, senior editor at *Canadian Running* magazine, and columnist for the *Globe and Mail*. He holds a master's in journalism from Columbia and a Ph.D. in physics from Cambridge, and he did his postdoctoral research with the U.S. National Security Agency. Please visit alexhutchinson.net.